Fundamental Constants in Mathematics & Physics

π, e, φ, $\zeta(3)$,
$\zeta(s)$, γ, δ, c,
Q, μ, α, G, h

Are they universal codes?

Shahin A. Shayan

Copyright © 2020 Shahin A. Shayan
Amazon Publishing

All rights reserved. No part of this publication may be reproduced, stored in a retrieval system, or transmitted, in any form or by any means, without prior permission in writing from the author.

"... nature is so constituted that it is possible logically to lay down such strongly determined laws that within these laws, only rationally determined constants occur."

Albert Einstein, Autobiographical Notes

"Today scientists recognize the Comma of Pythagoras, Pi, and the Golden Proportion, as well as the closely related Fibonacci sequence, are universal constants that describe complex patterns in astronomy, music, and physics."

Jonathan Black, Mark Booth

"Since only a narrow range of the allowed values for, say, the fine structure constant will permit observers to exist in the universe; we must find ourselves in the narrow range of possibilities which permit them, no matter how improbable they are. We must ask for the conditional probability of observing constants to take particular ranges, given that other features of the universe, like its age, satisfy necessary conditions for life."

John D. Barrow, The Constants of Nature: The Numbers That Encode the Deepest Secrets of the Universe

"Most circular and alternating growth patterns in mathematics and physics will have the Golden ratio etched in them. The existence of the Golden ratio in the dynamic equation of a system can be attributed to a scaled version of the unit rate of circular, alternating, or repetitive behavior (due to the effects of π) in combination with a continuous growth (due to the effects of e). If we can find a basic equation that relates the Golden ratio to the numbers e, and π, we can be certain of this type of dynamics. Surprisingly enough, there is such an equation."

"Fundamental constants in mathematics and physics are not derived; they seem to be the properties of and etched in the fabrics of the mathematical and physical world. They show up in most of our equations, in macro astronomical observations, micro quantum mechanical measurements, and they allow us the predictability, consistency, regularity, and continuity of understanding of the dynamics of our mathematical and physical world. What is so particular about these numbers? Are they hardcoded in the fabric structure of our universe? Are they the constant numbers needed for our universe to be the way it is? Were they required and set for the initial conditions for the big bang to start? Are they set and fine-tuned to be as they are? It seems like so."

From This Book

CONTENTS

Acknowledgments	i
Introduction	2
Constant Numbers in Mathematics	7
Constant Numbers in Physics	53
Final Word	92
Appendix A	105
About the Author	110

To my family, especially to my son Shahed, who helped me formalize the basic concepts in this book,
&
to all those that have contributed, sacrificed, and promoted the accumulation of knowledge throughout history.

INTRODUCTION

The world is dynamic and continuously changing at the Micro, Meso, and Macro levels. We continuously analyze these changes at the smallest particle levels, such as strings, fields, quarks, electrons, protons, etc. up to atomic, molecular, cellular, animal, human, global, planet, star, galactic, and cosmological levels. Everything moves, changes, and revolves due to the effects of Gravitational, Electromagnetic, Strong, and Weak nuclear forces acting on matter and energy through the Space-Time continuum[1]. We try to understand and analyze all changes utilizing numbers, mathematics, and different cause and effect relationships through the equations of nature[2], which can be simple, static, or in the form of differential equations with constant and varying parts. The numbers are fundamental tools for measurements and can be real (rational or irrational) or imaginary.

Examples of cause and effect relationships and equations of nature include:

---

[1] The space-time continuum is the notion that sees the three-dimensional space to be entangled and interacted with the concept of time. It does not see the three-dimensional space as an independent and untangled feature from the parameter of time (as Newton proposed).

[2] These relationships are described through the four fundamental forces of Gravity, Electromagnetic, Strong, and Weak (the last three have been unified through a Standard Model of forces in nature, which assumes that the root of all three forces are the same) plus some other important physical relationships such, energy mass equivalency or energy frequency relationship for wavelike particles and etc.

- *Plank-Einstein Equation*, which states that a small particle moving in a wavelike manner with frequency f, will always have an energy equal to E through the following relationship:

$$E = h \times f$$

Here, **h** is a constant proportionality number with a dimension of joules times seconds, called the Plank constant. The calculated value of which is[3]:

$$6.6260\ldots \times 10^{-34} \ (j.s)$$

- *Mass, Energy equivalence*, which states that there is an equivalency relationship between the mass m of an object and the equivalent energy captured in that mass. The equation is as follows:

$$E = m \times C^2$$

Here, **C** is a constant proportionality number with the dimension of meters per second, called the speed of light. The calculated value for this constant number is:

$$2.9979\ldots \times 10^8 \ (m.s^{-1})$$

- *Gravitational force equation*, which states that there is an equivalency relationship between the attractive gravitational force of two bodies of mass $m1$, $m2$, and the inverse of the distance r squared between them. The equation is as follows:

$$F = G \times \frac{m_1 \times m_2}{r^2}$$

Here, **G** is a constant proportionality number with a dimension of

---

[3] Units in here are: j = Joule, s = Second, m = Meter, kg = Kilogram, N = Newton, C = Charge unit.

meters cube times the inverse of kilogram and seconds square, called the gravitational constant. The calculated value for this constant number is:

$$6.6740\ldots \times 10^{-11} \; (m^3 . kg^{-1} . s^{-2})$$

- *Electrostatic force or Coulomb's equation*, which states that there is an equivalency relationship between the electrostatic force of two charges *q1* and *q2* and the inverse of the distance *r* squared between them. The equation is as follows:

$$F = k \times \frac{q_1 \times q_2}{r^2}$$

Here, **k** is a constant proportionality number with a dimension of newton times meters squared times the inverse of charge square, called the electric force constant or Coulomb's constant. The calculated value for this constant number is:

$$8.9875\ldots \times 10^9 \; (N.m^2.C^{-2})$$

We can show more examples and find out how many more constant numbers in nature have been detected[4] and measured. An essential feature of these constants is that they do not change when the variable parameters in their corresponding equations change. In other words, they are invariant and independent from the varying parameters in their

---

[4] For example in the Standard Model of elementary particles which unifies the three of nature's fundamental forces of Electromagnetic, Strong, and Weak, we can define up to 25 constants for the mass of Fermions including six Quarks (up, down, charm, strange, top, and bottom), and six Leptons (electron, electron neutrino, muon, muon neutrino, tau, and tau neutrino) plus six force generating particles of W+, W-, and Z Bosons (for weak force), Photons (for electromagnetic force), Gluons (for strong force), and Higgs (for mass generation). To mathematically construct the complicated Standard Model, we need to define 25 constant parameters. This is a topic beyond the concepts we would like to analyze in this book and we will refer the interested readers for simple descriptions presented in Appendix A, Wikipedia and other sources. https://en.wikipedia.org/wiki/Standard_Model

relevant equations. It would also be interesting to know if these constants are invariant through time. Science assumes that they are, but this has not yet been proven.

Nature's constant numbers can be dimensionless, meaning that they can be numbers representing orders or ratios. For example, the ratio of proton to electron mass has been constant and calculated to be around 1836 times. They can also represent a dimensionless feature of a geometric structure, such as a circle. The number Pi (approximately equals to 3.141....) is a dimensionless constant number that is the ratio of circumference to the diameter of any perfect circle in the universe. They can also be a universal feature of all dynamically changing systems moving towards chaotic behavior. The Feigenbaum constant number (equal to 4.669....) is the limiting ratio of each bifurcation interval when you have period-doubling in dynamic systems moving towards chaotic behavior. Finally, they can be Euler's constant number 2 for the difference between the vertices plus the faces minus the edges of any Platonic solid such as tetrahedron, dodecahedron, cube, etc.

Even though we have discovered and measured the physical constants, we still do not have any understanding of why they have their specific values. It would be ideal to have equations of nature without any constant numbers, but that has not yet happened.

It seems as if the mathematical and physical constants carry secret codes and play a vital role in the original design and required initial conditions of our universe. They are numbers that exist independent of our existence and have a reality of their own. We have been lucky to find and measure them through mathematical or physical relationships, and most probably, there are more of them that will be found.

In this book, we will first explain the concept of numbers and analyze important constant numbers in mathematics. Next, we will try to understand the critical constant numbers in physics and find out which ones are the more fundamental ones. In the final chapter, we deal with the question of, is mathematics invented or discovered?

All chapters include a total of thirty questions with suggested solutions. The book is for interested readers with some basic backgrounds in science and calculus. It is meant to be a general discussion of the topic of independence, stability, and invariance of the fundamental constants in mathematics and physics and to explain their role in understanding universal dynamics.

CONSTANT NUMBERS IN MATHEMATICS

What are the numbers? Are numbers an abstract concept, or are they real? Are they creations of our minds, or do they truly exist without a need for our observation? Are they independent or based on our experiences? These are essential questions that have occupied many mathematical philosophers throughout ages, starting from Plato up to now.

We recognize, analyze, and understand the world around us through patterns, symmetries, and the observed cause and effect relationships. To be able to measure, calculate, and mimic these patterns, symmetries, and cause and effect relationships, we need measurements. To measure and count, we need numbers.

Numbers are abstract objects that represent units for counting and measurement, and they are shown by symbols called numerals. For example, number 2 represents two units, and it is shown by the numeral 2 in the English language. The abstract concept for number 2 is similar in all languages, but spelled differently and represented by different numerals. The following table compares different numerals in some selected languages[5].

---

[5] The table was put together by Liza Gonzalez, and it is on: https://www.pinterest.com/pin/487655465874257156/.

	0	1	2	3	4	5	6	7	8	9
Arabic	٠	١	٢	٣	٤	٥	٦	٧	٨	٩
Bengali	০	১	২	৩	৪	৫	৬	৭	৮	৯
Chinese (simple)	〇	一	二	三	四	五	六	七	八	九
Chinese (complex)	零	壹	貳	參	肆	伍	陸	柒	捌	玖
Chinese 花碼 (huā mǎ)	〇	〡	〢	〣	〤	〥	〦	〧	〨	〩
Devanagari	०	१	२	३	४	५	६	७	८	९
Ethiopic		፩	፪	፫	፬	፭	፮	፯	፰	፱
Gujarati	૦	૧	૨	૩	૪	૫	૬	૭	૮	૯
Gurmukhi	੦	੧	੨	੩	੪	੫	੬	੭	੮	੯
Kannada	೦	೧	೨	೩	೪	೫	೬	೭	೮	೯
Khmer	០	១	២	៣	៤	៥	៦	៧	៨	៩
Lao	໐	໑	໒	໓	໔	໕	໖	໗	໘	໙
Limbu	᥆	᥇	᥈	᥉	᥊	᥋	᥌	᥍	᥎	᥏
Malayalam	൦	൧	൨	൩	൪	൫	൬	൭	൮	൯
Mongolian	᠐	᠑	᠒	᠓	᠔	᠕	᠖	᠗	᠘	᠙
Myanmar	၀	၁	၂	၃	၄	၅	၆	၇	၈	၉
Oriya	୦	୧	୨	୩	୪	୫	୬	୭	୮	୯
Tamil	௦	௧	௨	௩	௪	௫	௬	௭	௮	௯
Telugu	౦	౧	౨	౩	౪	౫	౬	౭	౮	౯
Thai	๐	๑	๒	๓	๔	๕	๖	๗	๘	๙
Tibetan	༠	༡	༢	༣	༤	༥	༦	༧	༨	༩
Urdu	٠	١	٢	٣	۴	۵	۶	٧	٨	٩

Additional numerals

	10	20	30	40	100	1000	10000	10^8	10^{12}
Chinese (simple)	十	廿	卅	卌	百	千	万	亿	兆
Chinese (complex)	拾				佰	仟	萬	億	兆

A number system must be based on a defined unit for counting, such as the unit bases of 2, 6, 10, 12, or 60. Our current number system is structured around a unit base of 10. This means that the number 114 is counted based on the addition of powers of 10 and calculated as follows:

$$114 = (1 * 10^2) + (1 * 10^1) + (4 * 10^0) = (1, 1, 4)_{base\ 10}$$

And, $(1, 1, 4)_{base\ 10}$ is usually shown as $114_{base\ 10}$.

Similarly, if we use a number system based on 2 (used in electronics and computer systems and known as the binary system of 0 and 1), the number 114 would look like:

$$114 = (1 * 2^6) + (1 * 2^5) + (1 * 2^4) + (0 * 2^3) + (0 * 2^2)$$
$$+ (1 * 2^1) + (0 * 2^0) = (1, 1, 1, 0, 0, 1, 0)_{base\ 2}$$

Where,

$(1, 1, 1, 0, 0, 1, 0)_{base\ 2}$ is usually shown as $1110010_{base\ 2}$.

And in a number system based on 12, the number 114 would be:

$$114 = (9 * 12^1) + (6 * 12^0) = (9, 6)_{base\ 12}$$

Where,

$(9, 6)_{base\ 12}$ is usually shown as $96_{base\ 12}$.

For base 60, we get:

$$114 = (1 * 60^1) + (54 * 60^0) = (1, 54)_{base\ 60}$$

Where,

$(1, 54)_{base\ 60}$ is usually shown as $154_{base\ 60}$.

We can, therefore, show number 114 in 4 different bases, as follows;

$$114_{base\ 10} = 1110010_{base\ 2} = 96_{base\ 12} = 154_{base\ 60}$$

The table below shows the numerals as they were used around 5,000 years ago by the Sumerian civilization[6]. In this civilization, the number system was based on units of sixty[7]. This meant that the counting system was based on units or multiples of 60.

𒁹	1	𒈫	2	𒐈	3	𒃻	4
𒐉	5	𒐊	6	𒐋	7	𒐌	8
𒐍	9	𒌋	10	𒌋𒁹	11	𒌋𒈫	12
𒌋𒐈	13	𒌋𒃻	14	𒌋𒐉	15	𒌋𒐊	16
𒌋𒐋	17	𒌋𒐌	18	𒌋𒐍	19	𒎙	20
𒌍	30	𒐏	40	𒐐	50	𒁹	60

Current theories of numbers are more abstract and are based on modern concepts of Set Theory[8], which assumes that the number system

---

[6] Sumerians are among the earliest known civilizations with writing, mathematics, astronomy and independent city-state legal systems. They were settled as early as 5,500 BC in the southern Mesopotamia region currently known as southern Iraq and south west of Iran. Their early writings in Cuneiform were recorded between the years 3,500 and 3,000 BC.

[7] Number 60 is the total number of joints in our fingers and feet. In Sumerian period and even now in some regions of the world, joints are used for counting and hence a total of 60 will be considered as one set, unit, round or base for counting. For example 72 horses will be shown as one complete unit or base of 60 plus 12 more and will be shown as $(1,12)_{base\ 60}$ or $112_{base\ 60}$. Using the above table, in Sumerian era, counting 72 horses would have been shown using the numeral for 60 first and 12 to its right. Our current calendar system, time, clocks and geometry are also based on the base units of 60. We have 60 minutes in an hour, 60 seconds in a minute and etc.

[8] Set Theory is known to be the fundamental theory in mathematics. Classical set theory is understood intuitively through defining a set as a collection of objects. For example the set of plates in a kitchen called set P, includes all plates in the kitchen represented by P = {plate1, plate2, plate3, ...}, and a set of forks called set F, includes all forks

is made up of a collection of subsets of numbers (or sets of number subsets). Simply put, it states that number zero is a set without a member called the null or the empty set. It then defines number one as a set that only includes the null set. Number two consists of the null and number one sets. Number three consists of the null, number one, and number two sets, and so on. We can show the sequence of numbers created using set theory as follows:

Number	Set Notation	
0	= { }	= ∅ null set
1	= { 0 }	= {∅}
2	= { 0, 1 }	= { ∅, {∅}}
3	= { 0, 1, 2 }	= { ∅, {∅}, {∅, {∅}}}
4	= { 0, 1, 2, 3 }	= { ∅, {∅}, {∅, {∅}}, {∅, {∅}, {∅, {∅}}}}
.		
.		

Therefore the modern number theory based on sets, assumes that the number system is a set of number sets.

There are also other theories of numbers available. Here, the basic definition that numbers are objects that represent units for counting and measuring will suffice. We want to know how numbers are classified. The classification of numbers is essential for the analysis of their properties and how they show up in nature.

A general and basic classification of numbers can be shown as follows:

- *Natural number set,* represented by the letter **N**
  This is the set of numbers that we are all familiar with. Numbers 1, 2, 3, etc. or **N** = {1, 2, 3 …}.This set does not include the

---

represented by F = {fork1, fork2, fork3, …}, and etc. This simple definition of sets can quickly lead to logical paradoxes, which were presented by Bertrand Russell and Ernst Zermelo. To resolve these paradoxes, the Axiomatic approach called the ZF set theory was put together by Ernst Zermelo, Abraham Fraenkel using certain axioms and logical deductions. For further readings see Appendix A and https://en.wikipedia.org/wiki/Set_theory#Axiomatic_set_theory

number zero. Zero is not a number that comes naturally to us when measuring or counting. Zero means nothing, and therefore, it is meaningless when measuring or counting. It becomes an important number when we start more sophisticated applications of numbers in understanding nature.

- ***Whole number set,*** represented by the letter **N0**
  When we add number zero to the natural number set, we get the whole number set. Therefore the Whole number set includes 0, 1, 2, 3, etc. or **N0** = {0, 1, 2, 3 ...}.

- ***Integer number set,*** represented by the letter **Z**
  When we add the set of negative natural numbers such as -1, -2, -3, etc., to the whole number set, we get the integer number set or **Z** = {..., -3, -2, -1, 0, 1, 2, 3 ...}.

- ***Rational number set,*** represented by the letter **Q**
  If we add numbers, composed of the ratios of two integer numbers (not including zero as the denominator) resulting in numbers such as 1.11, 1.23, 2.01, to the Integer number set Z, we get the rational number set or **Q** = {..., -3, -2, -1, 0, 1, 2, 3, ...}. This set includes all numbers in between each integer that can be shown as a fraction of two integer numbers. Therefore between integer numbers, there are no other numbers, while between rational numbers, there are many numbers between every two consecutive integers.

- ***Irrational number set***, represented by the letter **I**
  This is a number set that includes all numbers that cannot be written as the exact ratios (or fractions) of two natural numbers (not including zero) and will have decimals that will never end. It includes numbers such as Pi or π = 3.1415...., Euler's constant or e = 2.7182..., Apery's constant or $\zeta(3)$ = 1.2020..., Golden ration or φ = 1.6180..., $\sqrt{2}$ = 1.4142 ..., $\sqrt{3}$ = 1.7320 ...,

$\sqrt{5} = 2.2360$ ... and etc., or $I = \{.. , \pi, e, \zeta(3), \varphi, \sqrt{2}, \sqrt{3}, \sqrt{5}, ..\}$.

- **Real number set,** represented by the letter **R**
  This is a number set that includes both the rational and irrational numbers, meaning $\mathbf{R} = \mathbf{Q} \cup \mathbf{I}$ or real number set is a union of the rational and irrational number sets or, $\mathbf{R} = \{\mathbf{Q}, \mathbf{I}\}$.

- **Complex number set,** represented by the letter **C**
  This is a number set that includes numbers such as $\sqrt{-1}$ or its multiples. This number is represented by the numeral i, and therefore all complex numbers will be represented by the general form $(a + bi)$. In here a and b are both real numbers, and i is the complex number equal to $\sqrt{-1}$. Complex numbers can be written in general form of $C = \{a + b\,i\}$. We can assume that all other numbers are special forms of complex numbers. For example:

  - ✓ If b = 0, then $\mathbf{C} = \{a\}$ where $a = \mathbf{R}$
  - ✓ If a = 0, then $\mathbf{C} = \{b\,i\}$ where $b = \mathbf{R}$

Based on the above definition, we can state that N is a subset of N0, N0 is a subset of Z, Z is a subset of Q, Q is a subset of R, and R is a subset of C.

We can show the categorization of numbers in the following table:

General Number Classifications		
Real (**R**)		Complex (**C**)
Rational (**Q**)	Irrational (**I**)	
Integer (**Z**)		
Whole (**N0**)		
Natural (**N**)		

An interesting feature of the natural numbers (**N**) is that we can categorize them into Prime (**P**) and Non-Prime or composite numbers. Prime numbers are larger than one and can only be divided by number one and themselves. Non-prime natural numbers can always be created from the multiplication of prime numbers. ***Prime numbers can be considered as the building blocks of the natural number system. Every non-prime natural number or composite number can be constructed from Prime numbers.***

Current number theory states that we have an infinite number of prime numbers as we have natural numbers. The first ten prime numbers are as follows:

$$\mathbf{P} = \{2, 3, 5, 7, 11, 13, 17, 19, 23, 29, ...\}$$

There has been extensive abstract and theoretical analysis made on the nature of prime numbers and the possible existing patterns that can be detected among them[9]. For example, it can be proven that when a prime number greater or equal to 5 is squared, the result will always be a multiple of 24 plus one or:

$$\mathbf{P}^2 = 24n + 1 \quad \text{for } \mathbf{P} \geq 5$$

We have this relationship because prime numbers are either a multiple of six minus one or six plus one. Also, prime numbers are always going to be an odd natural number. Some Prime numbers can be written as the multiplication of several complex and complex conjugate pairs in the complex plane. For example number 5 can be written as a complex multiplication of (2+i)*(2-i) or (-2+i)*(-2-i) or (1-2i)*(1+2i) or (-1-2i)*(-1+2i) and so on. There are many other interesting properties that prime numbers have that are beyond the scope of this book to discuss.

---

[9] See Dunham, William (1999). "Euler, the Master of us all". The Mathematical Association of America, Library of Congress Catalog Card Number 98-88271.

Irrational Numbers (I)

We need to analyze irrational numbers further. They play an essential role in the mystery and patterns that lie in nature and the number system. As described before, when we have a number system that includes all numbers that cannot be written as the exact ratios (or fractions) of two natural numbers (not including zero) and will have decimals that will never end, we have irrational numbers. Fractional or decimal parts that never end means that these numbers are inexact, and we can only get some approximate value for them. In other words, they seem to be inexact constants in the number system, and strangely enough, they show up in many areas in mathematics and physics.

The fractional part of these numbers can have repeating patterns or no patterns at all. The fewer patterns we observe in the fractional portion, the more irrational they become. Let us cover the most important irrational numbers and try to analyze and understand them as much as possible.

- **Square Root of Two**

$$\sqrt{2} = 1.4142\ldots$$

In a square with sides equal to one, the diagonal line through the middle will split the square into two half equal right triangles. Using the Pythagorean theorem, the length of this diagonal will always be equal to $\sqrt{2}$, an irrational or an inexact fractional number that is numerically equal to 1.4142…. with the remainder going to infinity. This is an extraordinary result! How could an exact square with sides equal to one, an area of 1 and perimeter of 4 have such an inexact diagonal value of 1.4142…? Well, that is how nature plays a trick on us. *We accept this strange numerical value for $\sqrt{2}$ and define it as a constant irrational number.* We do not have any other choice. There is a fascinating mathematical analysis called Continued Fraction Analysis[10], which is used to detect specific mathematical patterns in all fractional numbers, including irrational numbers. For the square root of two, we get:

$$\sqrt{2} = 1 + \cfrac{1}{2 + \cfrac{1}{2 + \cfrac{1}{2 + \cfrac{1}{2 + \cfrac{1}{...}}}}}$$

The pattern of number 2 shows up in the denominator and continuing until infinity is amazing. The results of Fractional Analysis is usually shown as the set of integer coefficients of the nested fraction and for $\sqrt{2}$, the set is equal to [1;2,2,2,2,2,2,2,2,2,...][11]. The value of $\sqrt{2}$ continues forever and cannot be calculated exactly. It has also been proven mathematically that this number is equal to some interesting mathematical relationships. They include (sign $\prod$ means repeated multiplication):

$$\sqrt{2} = \prod_{n=0}^{n=\infty} (1 + \frac{1}{4n+1})(1 - \frac{1}{4n+3})$$

Numerically, this equals to:

$$\sqrt{2} = \left(1 + \frac{1}{1}\right)\left(1 - \frac{1}{3}\right)\left(1 + \frac{1}{5}\right)\left(1 - \frac{1}{7}\right)....$$

Or;

---

[10] See Appendix A.
[11] Ibid.

$$\sqrt{2} = \frac{3}{2} - 2\left(\frac{1}{4} - \left(\frac{1}{4} - \left(\frac{1}{4} - \cdots\right)^2\right)^2\right)^2$$

Or;

$$\sqrt{2} = 2\ Cos\ 45 = 2\ Sin\ 45$$

It is incredible to see such interesting geometric and algebraic relationships between rational and irrational numbers. We will see this occur again and again in mathematics.

o **Square Root of Three**

$$\sqrt{3} = 1.7320\ \ldots$$

Another irrational or none ending fractional number that is numerically equal to 1.7320 ... with a remainder going all the way to infinity is the square root of three. Whenever we have a cube with sides equal to one, we get an exact surface area of 6 and volume of 1, and the diagonal will be equal to $\sqrt{3}$. Also if we have an equilateral triangle with each side equal to 2, the height will be equal to the irrational number $\sqrt{3}$. Again, the sides, area, and perimeter of this triangle are all can exactly be calculated, and the height is not. This is very strange! *We accept this peculiar value for $\sqrt{3}$ and define it as another constant irrational number*. The Fractional Analysis for the square root of three is:

$$\sqrt{3} = 1 + \cfrac{1}{1 + \cfrac{1}{2 + \cfrac{1}{1 + \cfrac{1}{2 + \cfrac{1}{\cdots}}}}}$$

The alternating pattern of numbers 1 and 2 showing up in the denominator and continuing until infinity, is amazing. The Fractional Analysis can be shown to result in a set of integer coefficients such as [1;1,2,1,2,1,2,1,2,1...]. This shows that the value of $\sqrt{3}$, continues forever with a pattern in the fractional part and is not exact. An interesting mathematical relationship for the number $\sqrt{3}$ is:

$$\sqrt{3} = 2 - 2\left(\frac{1}{2} - \left(\frac{1}{2} - \left(\frac{1}{2} - \cdots\right)^2\right)^2\right)^2$$

- ## Square Root of Five

$$\sqrt{5} = 2.2360\ldots$$

Another irrational or none ending fractional number is equal to $\sqrt{5} = 2.2360\ldots$. The hypotenuse of a right triangle with one side equal to 2 and the other equal to 1 is equal to the irrational number $\sqrt{5}$. We have another **constant irrational number**. The Fractional Analysis for the square root of five is:

$$\sqrt{5} = 2 + \cfrac{1}{4 + \cfrac{1}{4 + \cfrac{1}{4 + \cfrac{1}{4 + \cfrac{1}{\cdots}}}}}$$

We see a number 4 pattern showing up in the denominator and continuing until infinity. The Fractional Analysis can be shown to result in a set of integer coefficients such as [2;4,4,4,4,4,4,4,4,...]. There are some interesting algebraic and geometric relationships between the irrational number $\sqrt{5}$ and other irrational numbers such as Pi or $\pi = 3.1415\ldots$, Euler's constant or $e = 2.7182\ldots$, and the Golden ratio or $\varphi = 1.6180\ldots$, which we will see later. Some interesting mathematical relations are shown below.

$$\sqrt{5} = \frac{1}{\sin 18} - 1$$

Or;

$$\sqrt{5} = 2\varphi - 1$$

$$\sqrt{5} = 3 - 10\left(\frac{1}{5} + \left(\frac{1}{5} + \left(\frac{1}{5} + \cdots\right)^2\right)^2\right)^2$$

- **Square Root of other Prime Numbers**

$$\sqrt{7}, \sqrt{11}, \sqrt{13}, \sqrt{17}, \sqrt{19}, \sqrt{23}, \sqrt{29} \ldots$$

Mathematically it has been proven that the square roots of other prime numbers such as 7, 11, 13, 17, and 19, etc., are all irrational constant numbers. The following table shows the values for the square root of the first ten prime numbers.

Square Root of Prime Numbers				
Prime	Equation	Value	Continued Fraction Integer Coefficients	Repeating Patterns (Period)
2	$\sqrt{2}$	1.4142...	[1;2,2,2,2,2,2,2,2,2...]	2 (1)
3	$\sqrt{3}$	1.7320...	[1;1,2,1,2,1,2,1,2,1...]	1,2 (2)
5	$\sqrt{5}$	2.2360...	[2;4,4,4,4,4,4,4,4,4...]	4 (1)
7	$\sqrt{7}$	2.6457...	[2;1,1,1,4,1,1,1,4,1...]	1,1,1,4 (4)
11	$\sqrt{11}$	3.3166...	[3;3,6,3,6,3,6,3,6,3...]	3,6,3 (3)
13	$\sqrt{13}$	3.6055...	[3;1,1,1,1,6,1,1,1,1...]	1,1,1,1,6 (5)
17	$\sqrt{17}$	4.1231...	[4;8,8,8,8,8,8,8,8,8,...]	8 (1)
19	$\sqrt{19}$	4.3588...	[4;2,1,3,1,2,8,2,1,3,...]	2,1,3,1,2,8 (6)
23	$\sqrt{23}$	4.7958...	[4;1,3,1,8,1,3,1,8,1,...]	1,3,1,8 (4)
29	$\sqrt{29}$	5.3851...	[5;2,1,1,2,10,2,1,1,2,..]	2,1,1,2,10 (5)

- **Pi**

$$\pi = 3.1415\ldots$$

If there is a circle with a diameter equal to one, the ratio of its perimeter to the diameter (two exact known numbers) will be equal to π, an irrational or none ending fractional constant number that is equal to 3.1415 .... This is a significant number representing harmonic and circular patterns in nature. It shows up in many areas of physics at the particle up to the cosmological levels. How could an exact circle with an exact diameter generate such an inexact ratio regardless of its size? *π is an important constant irrational number that is dimensionless or independent from the dimensions and sizes involved*. The Fractional Analysis for the number π is as follows:

$$\pi = 3 + \cfrac{1}{7 + \cfrac{1}{15 + \cfrac{1}{1 + \cfrac{1}{292 + \cfrac{1}{\ldots}}}}}$$

We do not observe any particular patterns in the denominator, which continues until infinity. The Fractional Analysis can be shown to result in a set of integer coefficients such as [3;7,15,1,292,1,1,1,2,1,...]. The number π has been subject to much interesting analysis, due to its importance and repeated occurrences in mathematics and physics. Fascinating work has been done by people such as Leonhard Euler[12] on the number π, and some of the results are shown below.

$$\frac{\pi^2}{6} = \sum_{n=1}^{n=\infty} \frac{1}{n^2}$$

---

[12] https://en.wikipedia.org/wiki/Leonhard_Euler

Or,

$$\pi = \sqrt{6\sum_{n=1}^{n=\infty} \frac{1}{n^2}}$$

Or,

$$\pi^2 = 6\left(1 + \frac{1}{2^2} + \frac{1}{3^2} + \cdots + \frac{1}{n^2} + \cdots\right)$$

Also,

$$\pi = 4\left(1 - \frac{1}{3} + \frac{1}{5} - \frac{1}{7} + \frac{1}{9} - \frac{1}{11} + \frac{1}{13} - \cdots\right)$$

And in terms of nested square roots we get:

$$\frac{2}{\pi} = \left(\frac{\sqrt{2}}{2}\right)\left(\frac{\sqrt{2+\sqrt{2}}}{2}\right)\left(\frac{\sqrt{2+\sqrt{2+\sqrt{2}}}}{2}\right)$$

Or;

$$\pi = \int_{-1}^{1} \frac{dx}{\sqrt{1-x^2}}$$

Or;

$$\frac{\pi}{2} = \left(\frac{2}{1} + \frac{2}{3}\right)\left(\frac{4}{3} + \frac{4}{5}\right)\left(\frac{6}{5} + \frac{6}{7}\right)\left(\frac{8}{7} + \frac{8}{9}\right)\cdots$$

Another fractional version of π with unusual patterns in the numerators and the denominators is:

$$\pi = \cfrac{4}{1 + \cfrac{1^2}{2 + \cfrac{3^2}{2 + \cfrac{5^2}{2 + \cfrac{7^2}{2 + \cfrac{9^2}{\ldots}}}}}}$$

And another strange relationship is (*ln* stands for natural Logarithm[13]):

$$\pi = 2\,\frac{\ln\sqrt{-1}}{\sqrt{-1}} = 2\,\frac{\ln i}{i}$$

Or;

$$\pi = \lim_{n \to \infty} \frac{e^{2n} n!^2}{2n^{2n+1}}$$

There have been many interesting analyses made on the nature and methods of calculation for this important irrational dimensionless constant of nature. ***Whenever there is a circular, alternating, or repetitive behavior in nature, the number π shows up.***

- ### Euler's Constant

$$e = 2.7182\ldots$$

Another irrational dimensionless constant number that is as important as the number π is e. It represents a specific type of long term cumulative growth factor observed in many instances in nature. The value of this number is equal to 2.7182 … It is a significant number representing a particular type of relative rate

---

[13] Natural logarithm of a number is the power **b** to which the irrational Euler's constant or e = 2.7182…would have to be raised to, for the result to be equal to that number or ln e^b = **b**. This means that:

$$\ln e^1 = 1 \text{ or } \ln e^e = e.$$

of change observed in radioactive decay, temperature change, harmonic growth, growth of interest on investments, and especially when something is cumulatively growing on itself. The Fractional Analysis for the number e is as follows:

$$e = 2 + \cfrac{1}{1 + \cfrac{1}{2 + \cfrac{1}{1 + \cfrac{1}{1 + \cfrac{1}{...}}}}}$$

We observe a repeated pattern of (1, even number, 1) with an increasing even number in the denominator, which continues until infinity. The Fractional Analysis can be shown to result in a set of integer coefficients such as [2;1,2,1,1,4,1,1,6,1,...]. A lot of analysis has also been done on the properties of this constant number. Some interesting results are as follows.

$$e = \sum_{n=0}^{n=\infty} \frac{1}{n!}$$

And,

$$e = \lim_{n \to \infty} \left(1 + \frac{1}{n}\right)^n$$

Moreover, other magical relationships are:

$$e = \lim_{n \to \infty} \frac{n}{\sqrt[n]{n!}}$$

The famous Euler's equation[14] which relates two significant

---

[14] A noted American 19th-century philosopher, mathematician, and a professor at Harvard University, Benjamin Pierce stated; It is absolutely paradoxical that we cannot understand it (Euler's Identity), and we don't know what it means, but we have proved it, and therefore we know it must be the truth.

irrational dimensionless numbers e, π, and the imaginary number $\sqrt{-1} = i$, to the numbers one (the basis of all numbers) and zero (the concept of nothingness), written by Euler to prove God's existence, is:

$$e^{i\pi} + 1 = 0$$

There have been many interesting analyses made on the nature and methods of calculation for the irrational constant number e. The general question that the results of basic mathematical operations such as addition, subtraction, division, and multiplication between irrational numbers such as e and π, can result in new irrational numbers, has not yet been proven.

By using scaling ideas, every number can be considered as a scaled version of number 1 or the base unit of counting. Every circle can also be considered as a scaled version of a unit circle of radius 1. When it comes to the number e, every continuous rate of growth can be considered as a scaled version of the unit rate of growth equal to e. In other words, the growth rate of all cumulatively growing systems is scaled versions of the standard growth rate equal to e.

- **Apery's Constant**

$$\zeta(3) = 1.2020...$$

This irrational constant number is equal to $\zeta(3) = 1.2020...$ Moreover, it is the converged value of the following infinite sum:

$$\zeta(3) = \sum_{n=1}^{n=\infty} \frac{1}{n^3}$$

This number arises in several physical phenomena, such as quantum electrodynamics and Stefan-Boltzmann law. The Fractional Analysis for the number $\zeta(3)$ is as follows:

$$\zeta(3) = 1 + \cfrac{1}{4 + \cfrac{1}{1 + \cfrac{1}{18 + \cfrac{1}{1 + \cfrac{1}{...}}}}}$$

There are no particular patterns in the denominator, which continues until infinity. The Fractional Analysis can be shown to result in a set of integer coefficients such as [1;4,1,18,1,1,1,4,1,1,...]. The interesting mathematical relations for this number are shown below.

$$\zeta(3) = \frac{8}{7} \sum_{n=0}^{n=\infty} \frac{1}{(2n+1)^3}$$

Or;

$$\zeta(3) = \frac{7\pi^3}{180} - 2 \sum_{n=1}^{n=\infty} \frac{1}{n^3 \cdot (e^{2\pi n} - 1)}$$

Or;

$$\zeta(3) = \iiint \frac{1}{1 - xyz} dx\,dy\,dz$$

Or;

$$\zeta(3) = \frac{1}{2} \int_0^\infty \frac{x^2}{(e^x - 1)} dx$$

The Apery's constant is a particular form of the Zeta function[15] $\zeta(s)$, such that s = 3. It is noteworthy to know that the number π is equal to the square root of 6 times the Zeta function $\zeta(s)$ such that s = 2. The general relationship of Reimann Zeta function and

---

[15] Reimann Zeta Function in mathematics is a form of inverse infinite series that can converge to a particular number. It has an important role in the analytical number theory. For further information see :
https://en.wikipedia.org/wiki/Riemann_zeta_function.

Euler product and prime numbers is shown as follows:

$$\zeta(s) = \sum_{n=1}^{n=\infty} \frac{1}{n^s} = \prod_{p}^{\infty} \frac{1}{1-p^{-s}} \text{ for } s > 1 \text{ and } p = Prime$$

o **Euler-Mascheroni Constant**

$$\gamma = 0.5772 \dots$$

This irrational constant number is equal to $\gamma = 0.5772$ .... It is the converged value of the difference between the sum of the inverse of number n (or a harmonic series) and the natural logarithm of number n, as n goes to infinity. Mathematically it is shown as:

$$\gamma = \lim_{n \to \infty} \left(\sum_{n=1}^{n} \frac{1}{n} - \ln n \right)$$

Euler-Mascheroni number γ is computed as the summation of the difference between the two graphs of the inverse of n and the logarithm of n, as n goes to infinity. This number arises in several physical phenomena, such as Shannon entropy in Quantum information theory, Levy distribution, Feynman diagrams in quantum field theory, etc. The Fractional Analysis for the number γ is as follows:

$$\gamma = 0 + \cfrac{1}{1 + \cfrac{1}{1 + \cfrac{1}{2 + \cfrac{1}{1 + \cdots}}}}$$

There are no particular patterns in the denominator, which continues until infinity. The Fractional Analysis can be shown to result in a set of integer coefficients such as [0;1,1,2,1,2,1,4,3,13,...]. The interesting mathematical relations for this number are shown below.

$$\gamma = \sum_{n=1}^{n=\infty} (\frac{1}{n} - \ln(\frac{n+1}{n}))$$

Or;

$$\gamma = -\int_0^\infty e^{-x} \ln x \, dx$$

There is an interesting relationship between this number and prime numbers. It is the limit of the following equation (p_i stands for the *ith* prime number):

$$\gamma = \lim_{n \to \infty} (\ln n - \sum_{p_i \leq n} \frac{\ln p_i}{p_i - 1})$$

- **Feigenbaum Constant**[16]

$$\delta = 4.6692...$$

An interesting irrational constant number is the Feigenbaum number, which is approximately equal to $\delta = 4.6692$. It is the limiting value of the ratio of the difference between two bifurcation intervals for each period-doubling of a one-dimensional discrete logistic equation[17] of the form (nonlinear iterative equation):

$$x_{n+1} = a \, x_n (1 - x_n)$$

δ is a universal constant number for all functions approaching chaotic behavior through period-doubling, and it is found through the above logistic equation as follows[18]:

---

[16] Watch the YouTube videos on;
https://www.youtube.com/watch?v=ETrYE4MdoLQ and
https://www.youtube.com/watch?v=ovJcsL7vyrk
[17] https://en.wikipedia.org/wiki/Feigenbaum_constants
[18] The calculated value for the Feigenbaum Constant to 30 decimal places is:

$$\delta = \lim_{n \to \infty} \frac{a_{n-1} - a_{n-2}}{a_n - a_{n-1}} = 4.6692 \ldots$$

Feigenbaum's number is a fingerprint for detecting instability and chaotic behavior (dynamic behavior that is highly sensitive to initial conditions) in nature. The closer a system's behavior is to the Feigenbaum constant, the more chaotic or sensitive it is to the initial conditions. The Fractional Analysis for the Feigenbaum number is as follows:

$$\delta = 4 + \cfrac{1}{1 + \cfrac{1}{2 + \cfrac{1}{43 + \cfrac{1}{2 + \cdots}}}}$$

There are no particular patterns in the denominator, which continues until infinity. The Fractional Analysis can be shown to result in a set of integer coefficients such as [4;1,2,43,2,163,2,3,1,1,2,...]. An interesting approximate and relatively precise mathematical equation for this number is:

$$\delta = \pi + \tan^{-1}(e^\pi) = 4.6692 \ldots$$

- **Golden Ratio**

$$\varphi \text{ or } \varphi 1 = 1.6180\ldots$$

This number is known to be the most irrational dimensionless constant number that we have known so far. It is a number that represents a specific type of growth rate that follows a particular pattern, observed in nature. Let us say you have a flower with length a, and after one month, it grows to length (a + b). If this relationship holds continuously every month, you have the Golden ratio phenomena at work. One can easily see the

---

$\delta = 4.66920160910299067185320 3821578$

following quadratic equation at work:

$$\frac{a+b}{a} = \frac{a}{b} = \varphi$$

And,

$$\frac{a+b}{a} = \frac{a}{a} + \frac{b}{a} = 1 + \frac{1}{\varphi}$$

Hence,

$$1 + \frac{1}{\varphi} = \varphi$$

$$\varphi^2 - \varphi - 1 = 0$$

And,

$$\varphi = \frac{1 \pm \sqrt{5}}{2}$$

$\varphi = 1.6180 \ldots$, and $-0.6180 \ldots$, which is called the conjugate

Interestingly enough, the number $\varphi = 1.6180 \ldots$ is equal to the ratio of two successive Fibonacci numbers[19]. The golden ratio shows up in many areas of natural sciences. It has been

---

[19] **Fibonnaci numbers** in mathematics are a sequence of numbers created starting from 1 and adding 1 to get 2, next you add 2 to the previous number 1 to get 3 and keep adding 3 to the previous resulted number 2 to get 5 and so on. By going through this algorithm you end up getting the Fibonacci numbers or series all the way up to infinity. These numbers are 1,1,2,3,5,8,13,21,34,55,... . If a Fibonacci number is Fn then the ratio Fn+1when n gets large becomes equal to the Golden ratio φ. Similar to this approach we can get the **Tribonacci numbers** that are a sequence that start with three predetermined terms and each term afterwards is the sum of the preceding three terms. The first few Tribonacci numbers are 0,0,1,1,2,4,7,13,24,44, ...By going through this algorithm you end up getting the Tribonacci numbers or series all the way up to infinity.

associated with the Golden Rectangles, Golden Triangles, and various symmetries in nature and beautiful architectures throughout history. The Fractional Analysis for the Golden ratio is as follows:

$$\varphi = 1 + \cfrac{1}{1 + \cfrac{1}{1 + \cfrac{1}{1 + \cfrac{1}{1 + \cfrac{1}{\ldots}}}}}$$

The repeating pattern of ones continuing until infinity is the reason for calling this constant number the most irrational known so far. The Fractional Analysis can be shown to result in a set of integer coefficients such as [1;1,1,1,1,1,1,1,1,...]. Some other impressive mathematical results are as follows.

$$\varphi = \frac{13}{8} + \sum_{n=0}^{n=\infty} \frac{(-1)^{n+1}(2n+1)!}{4^{2n+3}n!\,(n+2)!}$$

Or in terms of nested Square roots[20];

---

[20] The concept of nested square roots for a number x can be proved as follows: say x = r, then (x −r) = 0, and (x − r)(x+ (r -1)) is still equal to 0. This means;

$$x^2 - x - r(r - 1) = 0$$

And,

$$x = \sqrt{r(r - 1) + x}$$

And if we substitute for x we get:

$$x = \sqrt{r(r - 1) + \sqrt{r(r - 1) + x}}$$

And so forth;

$$\varphi = \sqrt{1 + \sqrt{1 + \sqrt{1 + \sqrt{1 + \sqrt{1 + \cdots}}}}}$$

Or;

$$\varphi = 1 + (2 \cdot \sin \pi/10)$$

$$\varphi = 2 \cos \pi/5$$

Or;

$$\varphi = \lim_{n \to \infty} \frac{F_n + 1}{F_n}$$

Where F_n is the nth Fibonacci number, It is interesting to know that successive powers of φ satisfy the following equation:

$$\varphi^{m+1} = \varphi^m + \varphi^{m-1}$$

Most "circular and alternating cumulative growth patterns" in mathematics or physics, will have the Golden ratio etched in them. The existence of the Golden ratio in the dynamic equation of a system can be attributed to a scaled version of the unit rate of circular, alternating, or repetitive behavior (due to the effects of π) in combination with a continuous cumulative growth rate (due to the effects of e). If we can find a basic equation that relates the Golden ratio to the numbers e, and π, we can be certain of this type of dynamics. Surprisingly enough, there is such an equation, and it is as follows:

$$x = \sqrt{r(r-1) + \sqrt{r(r-1) + \sqrt{r(r-1) + \sqrt{r(r-1) + x}}}}$$

$$\varphi = e^{\frac{+i\pi}{5}} + e^{\frac{-i\pi}{5}}$$

This equation shows that the internal dynamics of the Golden ratio is driven by the two important transcendental numbers e, and π. We can also show other expressions, such as:

$$\varphi = \frac{\frac{\pi}{\pi} \pm \sqrt{\frac{5e}{e}}}{2}$$

And in Continued Fraction form:

$$\varphi = \sqrt{\frac{1}{2}(5+\sqrt{5})} - \cfrac{e^{\frac{-2\pi}{5}}}{1 + \cfrac{e^{-2\pi}}{1 + \cfrac{e^{-4\pi}}{1 + \cfrac{e^{-6\pi}}{1 + \cfrac{e^{-8\pi}}{\ldots}}}}}$$

- **Metallic Ratios**

$$\varphi_n$$

If instead of having a flower with length a, after one month growing to length (a + b), the flower's length was divided into n equal parts and would grow to length (na + b). If this type of growth holds continuously every month, you have the general form of Golden ratio called the Metallic ratio phenomena at work. One can easily see the following quadratic equation at work:

$$\frac{na+b}{a} = \frac{na}{a} + \frac{b}{a} = \frac{a}{b} = n + \frac{1}{\varphi_n} = \varphi_n$$

Hence,

$$\varphi_n^2 - n\varphi_n - 1 = 0$$

And;

$$\varphi_n = \frac{n \pm \sqrt{n^2 + 4}}{2}$$

When n = 1, we get a value of 1.6180..., for the Golden ratio. For n = 2, we get a value of 2.4142 for the Silver ratio, and for n = 3, we get a value of 3.3027 for the Bronze ratio, and so on. These are different patterns of the rate of growth seen in nature. *It seems that the ratios of constant irrational Metallic numbers are ingrained in the fabrics of nature's cumulative repetitive growth and dynamics, which are observed everywhere.* The Fractional Analysis for the general Metallic ratios is as follows:

$$\varphi_n = n + \cfrac{1}{n + \cfrac{1}{n + \cfrac{1}{n + \cfrac{1}{n + \cfrac{1}{n + \cfrac{1}{\ldots}}}}}}$$

The repeating pattern of number n continuing until infinity is the reason for the Metallic constant numbers to be irrational. Interestingly enough, the Fractional Analysis can be shown to result in a set of integer coefficients such as [n;n,n,n,n,n,n,n,n,n ...]. Some other impressive general mathematical results for Metallic ratios are as follows.

$$\varphi_n = \sqrt{\varphi_n(\varphi_n - 1) + \sqrt{\varphi_n(\varphi_n - 1) + \sqrt{\varphi_n(\varphi_n - 1) + \varphi_n}}}$$

For example, for the first nine Metallic numbers we get:

$$\varphi_0 = \sqrt{0 + \sqrt{0 + \sqrt{0 + 1}}} = 1$$

$$\varphi_1 = \sqrt{1 + \sqrt{1 + \sqrt{1 + 1.6180\ldots}}} = \mathbf{1.6180\ldots}$$

$$\varphi_2 = \sqrt{3.4142\ldots + \sqrt{3.4142\ldots + \sqrt{3.4142\ldots + 2.4142\ldots}}} = \mathbf{2.4142\ldots}$$

$$\varphi_3 = \sqrt{7.6056\ldots + \sqrt{7.6056\ldots + \sqrt{7.6056\ldots + 3.3027\ldots}}} = \mathbf{3.3027\ldots}$$

$$\varphi_4 = \sqrt{13.7082\ldots + \sqrt{13.7082\ldots + \sqrt{13.7082\ldots + 4.2360\ldots}}} = \mathbf{4.2360\ldots}$$

$$\varphi_5 = \sqrt{21.7703\ldots + \sqrt{21.7703\ldots + \sqrt{21.7703\ldots + 5.1925\ldots}}} = \mathbf{5.1925\ldots}$$

$$\varphi_6 = \sqrt{31.8114\ldots + \sqrt{31.8114\ldots + \sqrt{31.8114\ldots + 6.1622\ldots}}} = \mathbf{6.1622\ldots}$$

$$\varphi_7 = \sqrt{43.8403\ldots + \sqrt{43.8403\ldots + \sqrt{43.8403\ldots + 7.1400\ldots}}} = \mathbf{7.1400\ldots}$$

$$\varphi_8 = \sqrt{57.8617\ldots + \sqrt{57.8617\ldots + \sqrt{57.8617\ldots + 8.1231\ldots}}} = \mathbf{8.1231\ldots}$$

The following table shows the first 10 Metallic ratios. They are all irrational constant dimensionless numbers in mathematics and nature, representing different types of dynamic growth patterns.

Metallic Ratios

n	Notation	Ratio	Value	Continued Fraction Integer Coefficients[21]	name
0	φ_0	$\dfrac{0 \pm \sqrt{4}}{2}$	1	[1;0,0,0,0,...]	Unit
1	φ_1	$\dfrac{1 \pm \sqrt{5}}{2}$	1.6180...	[1;1,1,1,1,...]	Golden
2	φ_2	$\dfrac{2 \pm \sqrt{8}}{2}$	2.4142...	[2;2,2,2,2,...]	Silver
3	φ_3	$\dfrac{3 \pm \sqrt{13}}{2}$	3.3027...	[3;3,3,3,3,...]	Bronze
4	φ_4	$\dfrac{4 \pm \sqrt{20}}{2}$	4.2360...	[4;4,4,4,4,...]	-
5	φ_5	$\dfrac{5 \pm \sqrt{29}}{2}$	5.1925...	[5;5,5,5,5,...]	-
6	φ_6	$\dfrac{6 \pm \sqrt{40}}{2}$	6.1622...	[6;6,6,6,6,...]	-
7	φ_7	$\dfrac{7 \pm \sqrt{53}}{2}$	7.1400...	[7;7,7,7,7,...]	-
8	φ_8	$\dfrac{8 \pm \sqrt{68}}{2}$	8.1231...	[8;8,8,8,8,...]	-
9	φ_9	$\dfrac{9 \pm \sqrt{85}}{2}$	9.1097...	[9;9,9,9,9,...]	-
10	φ_{10}	$\dfrac{10 \pm \sqrt{104}}{2}$	10.0990...	[10;10,10,10,10,...]	-
n	φ_n	$\dfrac{n \pm \sqrt{n^2 + 4}}{2}$	-	[n;n,n,n,n ...]	-

Similar to the Golden ratio, most growth patterns in mathematics or physics, have one of the Metallic ratios etched in their behavior. The existence of Metallic ratios in the dynamic equations of a system can be attributed to a scaled version of circular or repetitive behavior (due to the effects of π) in combination with a continuous cumulative growth dynamics (due to the effects of e). It is fascinating to see the existence of

---

[21] For all Metallic Ratios, the Fractional Analysis leads to Integer Coefficient periodicity of one.

patterns and structures, for constant dimensionless numbers that are considered irrational, non-ending, with "Continued Fraction Integer Coefficients" of the periodicity of one, playing such essential roles in the growth patterns, dynamic, and physical behaviors in nature.

We have summarized the results of our analysis for each irrational number, and they are shown in the following table:

Summary of the reviewed Irrational Numbers			
Number	Value	Continued Fraction Integer Coefficients	Repeating Patterns (Period)
$\sqrt{2}$	1.4142...	[1;2,2,2,2,2,2,2,2,2...]	2 (1)
$\sqrt{3}$	1.7320...	[1;1,2,1,2,1,2,1,2,1...]	1,2 (2)
$\sqrt{5}$	2.2360...	[2;4,4,4,4,4,4,4,4,4...]	4 (1)
$\sqrt{7}$	2.6457...	[2;1,1,1,4,1,1,1,4,1...]	1,1,1,4 (4)
$\sqrt{11}$	3.3166...	[3;3,6,3,6,3,6,3,6,3...]	3,6,3 (3)
$\sqrt{13}$	3.6055...	[3;1,1,1,1,6,1,1,1,1...]	1,1,1,1,6 (5)
$\sqrt{17}$	4.1231...	[4;8,8,8,8,8,8,8,8,8,...]	8 (1)
$\sqrt{19}$	4.3588...	[4;2,1,3,1,2,8,2,1,3,...]	2,1,3,1,2,8 (6)
$\sqrt{23}$	4.7958...	[4;1,3,1,8,1,3,1,8,1,...]	1,3,1,8 (4)
$\sqrt{29}$	5.3851...	[5;2,1,1,2,10,2,1,1,2,..]	2,1,1,2,10 (5)
π	3.1415...	[3;7,15,1,292,1,1,1,2,1,...]	None
e	2.7182...	[2;1,2,1,1,4,1,1,6,1,...]	1,1,even+2 (3)
ζ(3)	1.2020...	[1;4,1,18,1,1,1,4,1,1,...]	None
γ	0.5772...	[0;1,1,2,1,2,1,4,3,13,...]	None
δ	4.6692...	[4;1,2,43,2,163,2,3,1,1,2,...]	None
φ_1	1.6180...	[1;1,1,1,1,1,1,1,1,1,...]	1 (1)
φ_2	2.4142...	[2;2,2,2,2,2,2,2,2,2,...]	2 (1)
φ_3	3.3027...	[3;3,3,3,3,3,3,3,3,3,...]	3 (1)
φ_n	-	[n;n,n,n,n,n,n,n,n,n,...]	n (1)

A summary of the interesting observations and interrelationships among the reviewed dimensionless irrational constant numbers are:

- *Even though irrational constant numbers are not exact and have fractional parts that continue until infinity, when analyzed from a continued fraction, or other mathematical perspectives, they show beautiful mathematical patterns, relationships, and symmetries that are astounding. It is not clear how and why such exact patterns, relationships, or symmetries exist among these non-exact numbers.*

- *Numbers π, $\zeta(3)$, γ and δ, have no repeating patterns or periodicity in their continued fraction integer coefficients, which makes it difficult to calculate their values with high precision.*

- *Metallic Ratios φ_n, including the Golden Ratio φ_1 and Square roots of 17, 5, and 2 all have one repeating pattern or periodicity in their continued fraction integer coefficients. These numbers have beautiful symmetries ingrained in their calculations but are difficult to calculate with high precision.*

- *Eight other numbers, including square roots of 3, 7, 11, 13, 19, 23, and 29 with e show repeating patterns or periodicity of 3 or more numbers in their continued fraction integer coefficients. This has made it relatively less difficult to calculate their values with high precision.*

- *The following eleven interrelationships between the analyzed dimensionless irrational numbers are interesting to consider (many other relationships exist that have not been listed here):*

$$\sqrt{2} = 2\cos\left(\frac{\pi}{4}\right) = 2\sin\left(\frac{\pi}{4}\right) = \varphi_2 - 1$$

$$\sqrt{5} = \frac{1}{Sin(\frac{\pi}{10})} - 1 = 2\,\varphi_1 - 1$$

$$\varphi_1 = 2\,Cos(\frac{\pi}{5}) = e^{\frac{+i\pi}{5}} + e^{\frac{-i\pi}{5}} = \frac{1\pm\sqrt{5}}{2}$$

$$\varphi_2 = (2\,Cos(\frac{\pi}{4})) + 1 = e^{\frac{+i\pi}{4}} + e^{\frac{-i\pi}{4}} + 1 = \sqrt{2} + 1$$

$$\pi = \lim_{n\to\infty} \frac{e^{2n} n!^2}{2n^{2n+1}} = \sqrt{6 \sum_{n=1}^{n=\infty} \frac{1}{n^2}} = \sqrt{6\zeta(2)}$$

$$\frac{2}{\pi} = \left(\frac{\sqrt{2}}{2}\right)\left(\frac{\sqrt{2+\sqrt{2}}}{2}\right)\left(\frac{\sqrt{2+\sqrt{2+\sqrt{2}}}}{2}\right)\ldots$$

$$e^{i\pi} = -1 = i^2$$

$$\zeta(3) = \frac{1}{2}\int_0^\infty \frac{x^2}{(e^x-1)}dx = \frac{7\pi^3}{180} - 2\sum_{n=1}^{n=\infty} \frac{1}{n^3.(e^{2\pi n}-1)}$$

$$\gamma = -\int_0^\infty e^{-x}\ln x\,dx = \sum_{n=1}^{n=\infty}\left(\frac{1}{n} - \ln\left(\frac{n+1}{n}\right)\right)$$

$$\delta = \pi + \tan^{-1}(e^\pi)$$

$$\sqrt[i]{i} = i^{-i} = e^{\frac{\pi}{2}}$$

The existence of these mathematical interrelationships among dimensionless irrational constants, and complex numbers, forces us to explore the thought that there are probably more intricate underlying mathematical structures or designs present that we are not aware of and should search for. How could it be that such none exact constant irrational numbers end up having such surprising and precise mathematical interrelationships or patterns? The following matrix shows the existing interrelationships (+ sign) among the important constant dimensionless irrational numbers, analyzed so far.

	$\sqrt{2}$	$\sqrt{5}$	φ_1	φ_2	π	e	$\zeta(3)$	γ	δ	i
$\sqrt{2}$				+	+					
$\sqrt{5}$			+		+					
φ_1		+			+	+				
φ_2	+				+					
π	+	+	+	+		+	+		+	+
e			+		+		+	+	+	+
$\zeta(3)$					+	+				
γ						+				
δ					+	+				
i					+	+				
Total	2	2	3	2	8	6	2	1	2	2

The dimensionless irrational numbers π, and e that also happen to be transcendental[22], have the most interactions with the other irrational numbers. As stated before, <u>π shows up whenever we have harmonic, repetitive, or circular behaviors in nature</u>, and it is an important dimensionless irrational constant number that is independent of the dimensions and sizes involved. The number <u>e also shows up whenever something is growing on itself</u>, and all continually growing systems are scaled versions of a standard rate of number e. This is another important dimensionless irrational constant number that is independent of

---

[22] Transcendental numbers are irrational numbers that are not solutions to algebraic equations. For example, the irrational number $\sqrt{2}$ is the solution to the algebraic equation, $x^2 - 2 = 0$, therefore it is irrational and not a transcendental number. So is the Golden ratio φ_1 which is a solution to the equation, $x^2 - x - 1 = 0$, therefore it is an irrational but not a transcendental number. Irrational numbers such as π, e, π^{-1}, e^{-1} are not solutions to any algebraic equations, and therefore are considered to be transcendental. Therefore, all transcendental numbers are irrational and non-Algebraic. For further information see; https://en.wikipedia.org/wiki/Transcendental_number.

the dimensions and sizes involved. <u>The Golden ratio incorporates both features of the numbers π and e. It shows up whenever there is positive or negative repetitive cumulative spiral growth pattern,</u> such as seen in Whirlwinds, Whirlpools, mass movements around a black hole, and stars moving around the centers of galaxies.

Generally speaking, we still do not have a good understanding of the nature of these numbers and particularly the ones that are called Transcendentals[23], which are considered to be unique forms of irrational numbers. Cases such as π^π, π^e, π^{φ_1}, $\pi^{\sqrt{2}}$, $\pi^{\zeta(3)}$, π^γ and other irrational numbers added, multiplied, or raised to the power of similar irrational numbers are not very clear in terms of the resulting numbers being rational or irrational. These are areas of research that are still being explored by mathematicians. Hopefully, some clues and codes will be revealed through these searches.

To summarize, we will rewrite the table on the General Classification of Numbers with more details, as follows:

---

[23] Ibid

General Number Classifications in Mathematics

		Complex Numbers (C) $C = \{a + b\,i\}$ If $b = 0$, then $C = \{a\}$ where $a = \{R\}$ If $a = 0$, then $C = \{b\,i\}$ where $b = \{R\}$
Real Numbers (R) $R = \{Q, I\}$		
Rational Numbers (Q) $Q = \{\ldots, -a/b, \ldots, 0, \ldots, c/d, \ldots\}$ For a, b, c, d as integers & b and $d \neq 0$	Irrational Numbers (I) $I = \{\pi, e, \zeta(3), \delta, \varphi, \sqrt{2}, \sqrt{3}, \ldots\}$	
Integer Numbers (Z) $Z = \{\ldots, -3, -2, -1, 0, 1, 2, 3, \ldots\}$	Transcendental Numbers (T) $T = \{\pi, e, e^{\pi}, \sqrt{2}^{\sqrt{2}}, \sqrt{3}^{\sqrt{3}}, \ldots\}$	
Whole Numbers (N0) $N0 = \{0, 1, 2, 3, \ldots\}$		
Natural Numbers (N) $N = \{1, 2, 3, \ldots\}$		

The relationships between complex and irrational numbers through the stated equations and the way prime numbers behave as the fundamental building blocks for the number system, give us clues as to whether numbers are carrying some universal mathematical codes of nature. <u>By analyzing the relationship between the prime, irrational, and transcendental numbers, we could appreciate and understand the coded fundamental mathematical structure, symmetry, and pattern that have been etched in our universe.</u> Leonhard Euler[24] and Friedrich Gauss[25], the two fascinating mathematicians, were on this path and contributed greatly to this goal.

---

[24] **Leonhard Euler**. A Swiss mathematician, astronomer and physicist who lived from 1707 to 1783.
[25] **Johann Carl Friedrich Gauss**. A German mathematician and physicist who lived from 1777 to 1855.

Related Quotes

1. "He is unworthy of the name of a man who is ignorant of the fact that the diagonal of a square is incommensurable with its side."

   Plato
   https://www.azquotes.com/quotes/topics/numbers-and-math.html

2. "Physics is mathematical not because we know so much about the physical world, but because we know so little; it is only its mathematical properties that we can discover."

   Bertrand Russell
   https://www.azquotes.com/quotes/topics/numbers-and-math.html

3. "The Fibonacci Sequence turns out to be the key to understanding how nature designs... and is... a part of the same ubiquitous music of the spheres that build harmony into atoms, molecules, crystals, shells, suns, and galaxies and makes the universe sing."

   Guy Murchie, The Seven Mysteries of Life: An Exploration of Science and Philosophy
   https://www.goodreads.com/quotes/tag/golden-ratio

4. "The Golden Number is a mathematical definition of a proportional function which all of nature obeys, whether it is a mollusk shell, the leaves of plants, the proportions of the animal body, the human skeleton, or the ages of growth in man."

   R.A. Schwaller de Lubicz, Nature Word
   https://www.goodreads.com/quotes/tag/golden-ratio

5. "Thus, nature provides a system for proportioning the growth of plants that satisfies the three canons of architecture. All modules are

isotropic, and they are related to the whole structure of the plant through self-similar spirals proportioned by the golden mean."

Jay Kappraff, Connections: The Geometric Bridge Between Art and Science
https://www.goodreads.com/quotes/tag/golden-ratio

6. "The beauty of mathematics only shows itself to more patient followers."

Maryam Mirzakhani
https://www.brainyquote.com/search_results?q=Mathematical+Beauty

7. "The Golden Proportion, sometimes called the Divine Proportion, has come down to us from the beginning of creation. The harmony of this ancient proportion, built into the very structure of creation, can be unlocked with the 'key' ... 528, opening to us its marvelous beauty. Plato called it the most binding of all mathematical relations, and the key to the physics of the cosmos."

Bonnie Gaunt, Beginnings: The Sacred Design
https://www.goodreads.com/quotes/tag/golden-ratio

Questions

1. **Write the number 672 on the unit bases of 10, 12, and 60.**

   672 in the unit base of 10;

   $(6 * 10^2) + (7 * 10^1) + (2 * 10^0) = (6, 7, 2)_{base\ 10}$ or $672_{base\ 10}$

   672 in the unit base of 12;

   $(4 * 12^2) + (8 * 12^1) + (0 * 12^0) = (4, 8, 0)_{base\ 12}$ or $480_{base\ 12}$

   672 in the unit base of 60;

   $(11 * 60^1) + (12 * 60^0) = (11, 12)_{base\ 60}$ or $1112_{base\ 60}$

2. **Prove that for all Prime numbers:**

   $P^2 - 1 = 24n \quad or \quad P^2 = 24n + 1 \quad$ for $P \geq 5$

   Solution:

   We know that prime numbers are either a multiples of six minus one or six plus one. Therefore:

   - If Prime = 6n+1, n can either be odd (2r+1) or even (2r) or;

     $P = 6(2r + 1) + 1 = (12r + 7)$

$$P = 6(2r) + 1 = (12r + 1)$$

- If Prime $= 6n-1$, n can either be odd (2r+1) or even (2r) or;

$$P = 6(2r + 1) - 1 = (12r + 5)$$
$$P = 6(2r) - 1 = (12r - 1)$$

To see how these equation look when we have P^2 we get:

$$P^2 = (12r + 7)^2 = 24(6r^2 + 7r + 2) + 1$$
$$P^2 = (12r + 1)^2 = 24(6r^2 + r) + 1$$
$$P^2 = (12r + 5)^2 = 24(6r^2 + 5r + 1) + 1$$
$$P^2 = (12r - 1)^2 = 24(6r^2 - r) + 1$$

In all four cases, we proved that P^2 is a multiple of 24 plus one.

3. **Solve the following equation for x:**

$$4^x + 6^x = 9^x$$

Solution:

$$\frac{4^x}{4^x} + \frac{6^x}{4^x} = \frac{9^x}{4^x}$$

$$1 + (\frac{3}{2})^x = (\frac{3}{2})^{2x}$$

$$u^2 - u - 1 = 0 \text{ for } u = (\frac{3}{2})^x$$

$$u = (\frac{3}{2})^x = \frac{1 + \sqrt{5}}{2} = \varphi_1 \text{ and take the ln from both sides}$$

$$x = \frac{\ln u}{\ln(\frac{3}{2})} = \frac{\ln(\frac{1+\sqrt{5}}{2})}{\ln(\frac{3}{2})} = \frac{\ln \varphi_1}{\ln(\frac{3}{2})} \approx 1.187$$

4. **Show the Continued Fractional Analysis for the irrational number, square root of 31. Write the first 20 integer coefficients. Do you detect any patterns?**

   Solution:

   $$\sqrt{31} = 5 + \cfrac{1}{1 + \cfrac{1}{1 + \cfrac{1}{3 + \cfrac{1}{5 + \cfrac{1}{\ldots}}}}} = 5.5677\ldots$$

   [5;1,1,3,5,3,1,1,10,1,1,3,5,3,1,1,10,1,1,3,...]

   The repeating pattern is 1,1,3,5,3,1,1,and 10 with the periodicity of 8 exists.

5. **Show the Continued Fractional Analysis for the irrational number, square root of the square root of two. Write the first 20 integer coefficients. Do you detect any patterns?**

   Solution:

   $$\sqrt{\sqrt{2}} = 1 + \cfrac{1}{5 + \cfrac{1}{3 + \cfrac{1}{1 + \cfrac{1}{1 + \cfrac{1}{\ldots}}}}} = 1.18921\ldots$$

   [1;5,3,1,1,40,5,1,1,25,2,3,1,6,2,1,1,2,1,2,...]

   No repeating patterns are detected in the first twenty integer coefficients.

6. **Using any Software, calculate the Continued Fractions for the numbers, $(\varphi_1 \cdot \pi)$, and π^π. Write the first 20 integer coefficients. Do you detect any patterns in any of them?**

Solution:

$$(\varphi_1 \cdot \pi) = 5 + \cfrac{1}{12 + \cfrac{1}{53 + \cfrac{1}{2 + \cfrac{1}{14 + \cfrac{1}{\ldots}}}}} = 5.0832\ldots$$

[5;12,53,2,14,1,1,4,2,5,11,1,2,3,5,2,4,2,1,1,…]

No repeating patterns are detected in the first twenty integer coefficients. This is an irrational number.

$$\pi^\pi = 36 + \cfrac{1}{2 + \cfrac{1}{6 + \cfrac{1}{9 + \cfrac{1}{2 + \cfrac{1}{\ldots}}}}} = 36.4622\ldots$$

[36,2,6,9,2,1,2,5,1,1,6,2,1,291,1,38,50,1,2,5,…]

No repeating patterns are detected in the first twenty integer coefficients. This is an irrational number.

Based on the first 20 integer coefficients of the numbers analyzed, we can state that they are all ***dimensionless irrational constant numbers.*** The summary of the findings are shown in the following table:

	Summary of the Findings		
Number	Value	First Ten Integer Coefficients	Repeating Patterns (Period)
$(\varphi_1 \cdot \pi)$	5.0832…	[5;12,53,2,14,1,1,4,2,5,…]	None
π^π	36.4622…	[36;2,6,9,2,1,2,5,1,1, …]	None

7. **Using any Software, calculate the Continued Fractions for the numbers π^e, and $\pi^{\varphi 1}$. Write the first 20 integer coefficients. Do you detect any patterns in any of them?**

Solution:

$$\pi^e = 22 + \cfrac{1}{2 + \cfrac{1}{5 + \cfrac{1}{1 + \cfrac{1}{1 + \cfrac{1}{\ldots}}}}} = 22.4592\ldots$$

$$[22; 2, 5, 1, 1, 1, 1, 1, 3, 2, 1, 1, 3, 9, 15, 25, 1, 1, 5, 4, \ldots]$$

No repeating patterns are detected in the first twenty integer coefficients. This is an irrational number.

$$\pi^{\varphi 1} = 6 + \cfrac{1}{2 + \cfrac{1}{1 + \cfrac{1}{2 + \cfrac{1}{13 + \cfrac{1}{\ldots}}}}} = 6.3739\ldots$$

$$[6; 2, 1, 2, 13, 1, 6, 169, 20, 2, 1, 2, 35, 25, 5, 6, 1, 3, 19, 1, \ldots]$$

No repeating patterns are detected in the first twenty integer coefficients. This is an irrational number.

Based on the first 20 integer coefficients of the numbers analyzed, they are all *dimensionless irrational constant numbers*. The summary of the findings are shown in the following table:

Summary of the Findings			
Number	Value	First Ten Integer Coefficients	Repeating Patterns (Period)
π^e	22.4592...	[22;2,5,1,1,1,1,1,3,2, ...]	None
π^{φ_1}	6.3739...	[6;2,1,2,13,1,6,169,20,2,...]	None

8. Using any Software, calculate the Continued Fractions for the numbers $\pi^{\sqrt{2}}, e^{\frac{\pi}{2}}, \sqrt[\pi]{e}, \sqrt[\pi]{\pi}$ and δ^π. Write the first 20 integer coefficients. Do you detect any patterns in any of them?

Solution:

$$\pi^{\sqrt{2}} = 5 + \cfrac{1}{21 + \cfrac{1}{18 + \cfrac{1}{1 + \cfrac{1}{1 + \cfrac{1}{...}}}}} = 5.0475 ...$$

[5;21,18,1,1,2,1,16,1,5,1,9,15,2,1,1,1,1,10,1,...]

No repeating patterns are detected in the first twenty integer coefficients. This is an irrational number.

$$e^{\frac{\pi}{2}} = \sqrt[i]{i} = 4 + \cfrac{1}{1 + \cfrac{1}{4 + \cfrac{1}{3 + \cfrac{1}{1 + \cfrac{1}{...}}}}} = 4.8104 ...$$

[4;1,4,3,1,1,1,1,1,1,1,1,7,1,20,1,3,6,10,3,...]

No repeating patterns are detected in the first twenty integer coefficients. This is an irrational number.

$$\sqrt[e]{e} = e^{\frac{1}{e}} = 1 + \cfrac{1}{2 + \cfrac{1}{4 + \cfrac{1}{55 + \cfrac{1}{27 + \cfrac{1}{\cdots}}}}} = 1.4446\ldots$$

$$[1;2,4,55,27,1,1,16,9,3,2,8,3,2,1,1,4,1,9,6,\ldots]$$

No repeating patterns are detected in the first twenty integer coefficients. This is an irrational number.

$$\sqrt[\pi]{\pi} = \pi^{\frac{1}{\pi}} = 1 + \cfrac{1}{2 + \cfrac{1}{3 + \cfrac{1}{1 + \cfrac{1}{1 + \cfrac{1}{\cdots}}}}} = 1.4396\ldots$$

$$[1;2,3,1,1,1,3,1,1,3,2,1,6,3,4,2,1,14,1,1,\ldots]$$

No repeating patterns are detected in the first twenty integer coefficients. This is an irrational number.

$$\delta^{\pi} = 126 + \cfrac{1}{1 + \cfrac{1}{1 + \cfrac{1}{1 + \cfrac{1}{1 + \cfrac{1}{\cdots}}}}} = 126.6157\ldots$$

$$[126;1,1,1,1,1,1,15,1,2,1,24,9,2,1,13,2,14,3,3,\ldots]$$

No repeating patterns are detected in the first twenty integer coefficients. This is an irrational number.

Based on the first 20 integer coefficients of the numbers analyzed, they are all **dimensionless irrational constant numbers**. The summary of the findings are shown in the following table:

	Summary of the Findings		
Number	Value	First Ten Integer Coefficients	Repeating Patterns (Period)
$\pi^{\sqrt{2}}$	5.0475...	[5;21,18,1,1,2,1,16,1,5,...]	None
$\sqrt[i]{i} = e^{\frac{\pi}{2}}$	4.8104...	[4;1,4,3,1,1,1,1,1,1,,...]	None
$\sqrt[e]{e} = e^{\frac{1}{e}}$	1.4446...	[1;2,4,55,27,1,1,16,9,3,,...]	None
$\sqrt[\pi]{\pi} = \pi^{\frac{1}{\pi}}$	1.4396...	[1;2,3,1,1,1,3,1,1,3,,...]	None
8^{π}	126.6157...	[126;1,1,1,1,1,1,15,1,2,,...]	None

9. **Show the Continued Fractional Analysis for the complex number i^i. Write the first 20 integer coefficients. Do you detect any patterns? Does this look like an irrational number?**

Solution:

We know that;

$$i^i = e^{\frac{-\pi}{2}}$$

Then;

$$i^i = e^{\frac{-\pi}{2}} = 0 + \cfrac{1}{4 + \cfrac{1}{1 + \cfrac{1}{4 + \cfrac{1}{3 + \cfrac{1}{...}}}}} = 0.2078\ldots$$

[0;4,1,4,3,1,1,1,1,1,1,1,1,7,1,20,1,3,6,10,...]

No repeating patterns are detected in the first twenty integer coefficients. This complex number is a ***dimensionless irrational number***. Many complex numbers are irrational. This particular number is also transcendental because of Gelfond-Schneider Theorem[26].

10. Using any Software and the Gelfond-Schneider Theorem (GST)[27] for the numbers, $\varphi_1^{\varphi_1}, \sqrt{2}^{\sqrt{2}}, \sqrt{2}^{\varphi_1}, \sqrt{3}^{\sqrt{3}}, \sqrt{3}^{\varphi_1}, 2^{\sqrt{2}}$ and determine if they are transcendental or not.

Solution:

We used the Mathematica software to calculate the results in the following table.

Number	GST Conditions met	Value	First Ten Integer Coefficients	Repeating Patterns (Period)	Transcendental
$\varphi_1^{\varphi_1}$	Yes	2.1784...	[2;5,1,1,1,1,10,1,1,2, ...]	None	Yes
$\sqrt{2}^{\sqrt{2}}$	Yes	1.6325...	[1;1,1,1,2,1,1,2,2,1, ...]	None	Yes
$\sqrt{2}^{\varphi_1}$	Yes	1.7520...	[1;1,3,30,1,2,1,2,1,4, ...]	None	Yes
$\sqrt{3}^{\sqrt{3}}$	Yes	2.5894...	[2;1,1,2,3,2,1,2,10,2,...]	None	Yes
$\sqrt{3}^{\varphi_1}$	Yes	2.4322...	[2;2,3,5,2,1,18,128,7,2, ...]	None	Yes
$2^{\sqrt{2}}$	Yes	2.6651...	[2;1,1,1,72,3,4,1,3,2,...]	None	Yes

---

[26] Gelfond-Schneider Theorem (or GST) states that, if **A** and **B** are algebraic numbers (and not equal to zero or 1) and B is irrational then any value of A^B is transcendental.
[27] Ibid.

CONSTANT NUMBERS IN PHYSICS

Analyzing numbers in mathematics can get pretty abstract, and through this analysis, we were able to see a wide range of important inexact dimensionless constant numbers with interesting properties. Modern science and in particular physics is founded almost entirely on mathematics. In many instances, our efforts to understand the physical universe have directed us to develop or find new mathematics. One of the striking features of our universe is that its laws are mathematical. We do not know why, but that is how it is. In physics, the same mathematical dimensionless constant numbers such as π, e, γ, and ζ appear in addition to some new constant numbers that show relationships between universal parameters such as mass, energy, charge, etc.

Constant numbers in physics show relationships between sets of variables. They usually are conversion factors between the two sides of equations in nature. Without having any understanding of why they possess the specific values that they have, they are mostly dimensional. For example, in Einstein's equation of mass and energy equivalence we have[28]:

---

[28] A more general version of this equation is written in relativistic terms as follows:

$$E^2 = (pc)^2 + (m_0 c^2)^2$$

$$E = mc^2$$

Where E is the equivalent energy for rest mass m, the conversion factor is the square of the constant speed of light or c^2 with the dimension of meters per seconds, squared $(m.s^{-1})^2$. Also, in the relationship between the photon energy and its wave frequency which is;

$$E = h\nu$$

Where E is the photon energy with frequency ν, the conversion factor is the Plank's constant number or h with the dimension of joules times seconds $(j.s)$.

Similar to constants in mathematics, constants in physics can be dimensionless. For example, the Fine structure constant, which is approximately equal to 1/137 (which we will cover later) or Proton to Electron mass ratio, which is approximately equal to 1836, are dimensionless constant numbers in physics.

Nature is continually changing, and we try to understand how these changes are made. Stability, invariance, and permanence in the physical world is a feature that gives us a sense of beauty and understanding in terms of the direction of change and the way the future could look.

For us, Existence seems to reveal itself through changes in space and time[29]. If we can find permanence and stability in the equations or the structure of our universe, we get a sense of continued existence.

Invariance, permanence, resilience, and constancy in the universe can be observed in fundamental equations and important constant numbers that seem to be a part of the building blocks of the universal structure and design. Einstein believed that if we find a theory or equation that

---

E is the equivalent energy for a particle with momentum p, and rest mass $m0$ moving with the speed of v. At a speed $v = 0$, there is no momentum or $p = 0$. In this situation a particle with rest mass m_0, will have an energy due to its mass equal to:

$$E = \sqrt{(m_0 c^2)^2} = m_0 c^2$$

[29] **Shayan S. A. (2020).** "Existence, A way to see It." Independent Publisher, Amazon.

unifies all fundamental forces of nature defined as the Theory of Everything[30], we would have minimum constant numbers with their values precisely explained by the internal consistency of the theory. He also believed that these minimum constant numbers in nature would be divine inputs that are required over and above the Theory of Everything (laws of nature) and for the starting conditions of a unique universe.

We will explore some of the more critical fundamental laws of nature with their known constant numbers to pose essential questions.

- **Gravitational Constant**

$$G = 6.6740 \ldots \times 10^{-11} \ (m^3 \cdot kg^{-1} \cdot s^{-2})$$

This dimensional constant number shows the equivalency of gravitational force to the masses $m1$ and $m2$ of two objects and the inverse of their distance r squared. Or;

$$F = G * (\frac{m_1 * m_2}{r^2})$$

Or;

$$G = \frac{F}{(\frac{m_1 * m_2}{r^2})}$$

G is called the gravitational constant, which causes the force of gravity F to equate to the variable terms $m1$, $m2$, and r on the right-hand side of the equation. It is a constant proportionality number defined by Newton and calculated by many experiments through finding the value for the above ratio for G. The calculated value has been observed to be constant, and we have assumed that it is independent of the other variables. We have also assumed

---

[30] Theory of Everything or as it is sometimes called Unified Field Theory, is an all-encompassing, theoretical framework of physics that tries to explain and link together all the known physical laws in the universe. This theory unifies all of the fundamental interactions of nature including the strong, weak, electromagnetic, and gravitational interactions. Currently, there are no accepted and candidate for the Unified Theory of everything.

that the gravitational equation correctly defines the value of the force of gravity. We have also assumed that the force of gravity between two objects is only dependent on m1, m2, r, and the constant value G. These are important assumptions that should make us think. The calculated value for G has been:

$$6.6740 \ldots \times 10^{-11} \ (m^3.kg^{-1}.s^{-2})$$

Is **G** constant in all space-time situations such as highly curved spaces near black holes or at the time of the big bang? Is G constant through time, or is G (t)? Is G constant near the speed of light? Is G constant at quantum levels? Is G constant in other possible universes? In other words, is G truly constant, or is it constant at the levels of human measurements and experimental abilities? We do not know the accurate answer to these questions. We know this is a very important constant number in physics that shown up in many equations.

- **Magnetic Constant**

$$\mu_0 = 4\pi \times 10^{-7} \ (H.m^{-1})$$

It is also known as the permeability constant, the permeability of vacuum, permeability of free space, or vacuum permeability. This dimensional constant number (in units of Henry H per meter m), shows the equivalency relationship of magnetic force per unit length F_l, created by the current j carried in a wire, separated from another wire with distance r and the inverse of the square of their distance r. The equation is as follows:

$$F_l = \frac{\mu_0}{2\pi} * (\frac{j^2}{r})$$

Or;

$$\mu_0 = \frac{2\pi r F_l}{j^2}$$

Here, μ_0 is a fundamental constant number and a property of the vacuum or free space called the Permeability constant. It can also be interpreted as the vacuum allowance for the movement of the magnetic field. It is a property of free space or vacuum, and it is not a derived number. The larger this constant number is, the larger the magnetic force field will be (maybe in another universe). The calculated value is:

$$4\pi \times 10^{-7} \; (H.m^{-1}) \; or \; 12.5663 \ldots \times 10^{-7} \; (H.m^{-1})$$

Interestingly enough, the mathematical inexact dimensionless transcendental constant number π shows up in this equation. Another very interesting mathematical relationship exists between the magnetic constant μ_0, Plank's constant h, fine structure constant α, the electric charge of an electron e, electric constant or permittivity constant ε_0, and the speed of light in vacuum c (which all will be covered later). This relationship is a very unusual mathematical relationship among five very important constants of nature. It is shown below.

$$\mu_0 = \frac{2\alpha h}{e^2 c} = \frac{1}{\varepsilon_0 c^2}$$

Or;

$$\mu_0 = \frac{2\alpha h}{e^2 c} = \frac{2\pi r F_l}{I^2}$$

These are very interesting and peculiar relationships that show up in mathematics and physics. We are not sure why this interesting relationship among the constant numbers μ_0, α, h, e, c, ε_0, and π exist, and we are glad that they exist. These relationships could be a clue to a deeper relationship, design, or structure in between the known natural laws, equations, and fundamental forces of nature, that need to be explored.

Is $\mu 0$ constant in all space-time situations such as highly curved spaces near black holes or at the time of the big bang? Is $\mu 0$ constant through time? Is $\mu 0$ constant at quantum levels? Is $\mu 0$ constant in other possible universes? Is $\mu 0$ truly constant, or is it constant at the levels of human measurements and experimental abilities? We do not know the accurate answer to these questions either but have assumed that it is constant, and our current experiments have confirmed this assumption.

- **Electric Constant**

$$\varepsilon_0 = 8.8541 \ldots \times 10^{-12} \ (F.m^{-1})$$

This dimensional constant (in Faraday F per meter m) in physics is known as the permittivity constant, permittivity of free space, or the distributed capacitance of the vacuum. This number is related to Coulomb's constant k in the equivalency relationship between the electrostatic force between two charged particles $q1$ and $q2$ and the inverse of their distance r squared, through the equation:

$$F = k * \left(\frac{q_1 q_2}{r^2}\right) = \frac{1}{4\pi\varepsilon_0} * \left(\frac{q_1 q_2}{r^2}\right)$$

Or;

$$\varepsilon_0 = \frac{1}{4\pi k}$$

Here, $\varepsilon 0$ is a fundamental constant number called the Permittivity constant, which can be interpreted as the vacuum resistance for the movement of the electric field. It is a property of the free space or vacuum, and it is not a derived number. The larger this constant number is, the smaller the electric force field will be (maybe in another universe). The calculated value is:

$$8.8541... \times 10^{-12} \ (F.m^{-1})$$

A series of mathematical relationships exist between the electric constant $\varepsilon 0$, magnetic constant $\mu 0$, coulomb constant k, the fine structure constant α, plank constant h, the transcendental mathematical constant π, and the speed of light in vacuum c[31]. Again these are very unusual mathematical relationships among fundamental constants of nature. They are as follows:

$$\varepsilon_0 = \frac{1}{\mu_0 c^2} = \frac{1}{4\pi k} = \frac{e^2}{2\alpha hc}$$

These are other peculiar relationships that show up in mathematics and physics. We are not sure why they exist. Again, they might be clues to some hidden structures in nature that we are not aware of.

Is $\varepsilon 0$ constant in all space-time situations such as highly curved spaces near black holes or at the time of the big bang? Is $\varepsilon 0$ constant through time? Is $\varepsilon 0$ constant at quantum levels? Is $\varepsilon 0$ constant in other possible universes? Is $\varepsilon 0$ truly constant, or is it constant at the levels of human measurements and experimental

---

[31] It was the following equation:

$$c = \sqrt{\frac{1}{\varepsilon_0 \mu_0}}$$

that was used by James Clerk Maxwell the famous 19th century Scottish physicists and mathematician to determine the speed of electromagnetic fields in free space or vacuum and that light was an electromagnetic radiation. Einstein concluded that, due to the fact that the speed of electromagnetic field or light was determined by two fundamental electromagnetic constant properties of nature it had to be a fundamental constant and property in free space, vacuum and nature.

abilities? We have assumed that it is constant, and our current experiments have confirmed this assumption also.

- **Plank's Constant and Plank Scales**

$$h = 6.6260\ldots \times 10^{-34} \ (J.s)$$

Plank constant is a dimensional constant number (in *Joules* times *Seconds*) showing the equivalency of photon energy E with the photon wave frequency v or;

$$E = hv$$

Or;

$$h = \frac{E}{v}$$

Plank constant h causes the photon energy to equate to the photon frequency on the right-hand side of the equation. It is a constant proportionality number defined by Max Plank[32] and

---

[32] Max Plank, the great German physicist, utilized fascinating analysis and mathematical derivation to conclude that there are minimum scales or limits for Length, Time, Mass, Charge, and Temperature in nature that we can observe and calculate. These scales were derived from five critical constants of nature; the Speed of light "C," used in Relativity, Reduced Planks constant "H or Plank constant divided by two times Pi," used in Quantum Mechanics, Gravitational constant "G," used in Gravitational theories, Coulomb constant "K," used in Electricity and Magnetism, and Boltzmann constant "Kb," Used in statistical behavior of gases and laws of Thermodynamics. These constants are universal and constant through time. If they increase or decrease, the calculated scales will also change (remember the Fine-Tuning Principle). The equations and values calculated for the five important Plank scale limits are as follows:

Plank Length = $\sqrt{\frac{H*G}{C^3}}$ $1.6162\ldots \times 10^{-35}$ (*meter*)

Plank Time = $\sqrt{\frac{H*G}{C^5}}$ $5.3912\ldots \times 10^{-44}$ (*second*)

Plank Mass = $\sqrt{\frac{H*C}{G}}$ $2.1764\ldots \times 10^{-8}$ (*kilogram*)

calculated through the blackbody radiation experiments. Here we have assumed that the above equation correctly defines the value for the photon energy and has assumed that the photon energy E is only dependent on photon frequency v and the plank constant h. The calculated value for h has been:

$$6.62607015 \times 10^{-34} \; (j.s)$$

Based on Plank's constant and theoretical analysis, when dealing with Quantum Mechanical sizes, there are Plank scale limits for Length, Time, Mass, Charge, and Temperature to be considered that are a part of the fabric of the universe[33]. We consider them as limits of constant numbers in physics, and they are (H is equal to h divided by 2π):

Plank Length $= \sqrt{\dfrac{H*G}{C^3}}$ $1.6162\ldots \times 10^{-35} \; (meter)$

Plank Time $= \sqrt{\dfrac{H*G}{C^5}}$ $5.3912\ldots \times 10^{-44} \; (second)$

Plank Mass $= \sqrt{\dfrac{H*C}{G}}$ $2.1764\ldots \times 10^{-8} \; (kilogram)$

Plank Charge $= \sqrt{\dfrac{H*C}{K}}$ $1.8755\ldots \times 10^{-18} \; (coulomb)$

Plank Temperature $= \sqrt{\dfrac{H*C^5}{G*Kb^2}}$ $1.4167\ldots \times 10^{+32} \; (kelvin)$

There is also an equation that relates h to the fine structure constant α, inversely to the speed of electromagnetic field or

---

Plank Charge $= \sqrt{\dfrac{H*C}{K}}$ $1.8755\ldots \times 10^{-18} \; (coulombs)$

Plank Temperature $= \sqrt{\dfrac{H*C^5}{G*Kb^2}}$ $1.4167\ldots \times 10^{+32} \; (kelvin)$

[33] Ibid

wave (light) in vacuum c, the magnetic constant (permeability) or $\mu 0$ and the Josephson constant (related to the frequency of microwave radiation) or Kj, as follows:

$$h = \frac{8\alpha}{\mu_0 c K_j^2}$$

Is **h** constant in all space-time situations such as highly curved spaces near black holes or at the time of big bang? Is **h** constant through time? Is **h** constant near the speed of light? Is **h** constant in other possible universes? In other words, is **h** truly constant, or is it constant at the levels of human measurements and experimental abilities? We do not know the accurate answer to these questions. We know this is a fundamental constant in physics that shown up in many equations such as the Heisenberg's uncertainty principle[34]. Interestingly, in this principle, two important constants, *h,* and π (transcendental number) play an essential role.

- **Speed of Electromagnetic Field or Wave in Vacuum**

$$c = 2.99792458 \times 10^8 \; (m.s^{-1})$$

This is a dimensional constant number (in *Meters per Second*) calculated through Maxwell's equations which also shows up in

---

[34] Heisenberg Uncertainty Principle, states that in the quantum mechanical world, there is an uncertainty relationship that applies when defining the position and momentum of a particle that has wavelike behavior. This relationship is as follows:

$$\Delta x \times \Delta p_x \geq \frac{h}{4\pi}$$

where Δx is the uncertainty in the position of a particle and Δp_x is the uncertainty in the momentum of the same particle. Interestingly enough in this relationship two important constants h and π play an important role.

Einstein's equivalency equation of mass and energy as follows[35]:

$$E = mc^2$$

Where E is the equivalent energy for rest mass m, the conversion factor is the square of the constant speed of the electromagnetic field[36] or wave in a vacuum (this is also the speed of light, since light is known to be a form of electromagnetic wave) or c^2 with the dimension of meters per seconds, squared $(m.s^{-1})^2$. The constant speed of the electromagnetic field in vacuum c, when squared and multiplied by the rest mass m, will give us the equivalent energy for the rest mass m. Here we have assumed that the above equation correctly defines the value for the equivalent energy of rest mass m, and the energy E is only dependent on rest mass m and the square of the speed of the electromagnetic field or wave. The calculated value for c is:

$$2.99792458 \times 10^8 \ (m.s^{-1})$$

Is c constant in all space-time situations such as highly curved spaces near black holes or at the time of big bang? Is c constant through time, or is c is $c\ (t)$? Is c constant in other possible universes? In other words, is c truly constant, or is it constant at the levels of human measurements and experimental abilities? Through Maxwell's electromagnetic equations[37] and

---

[35] See footnote 28.
[36] In physics, fields are defined as any representation in free space that can create a physical effect or cause changes. Fields can have scalar, vector, or tensor values at each points in space. Example of fields include; Temperature fields that are considered scalar fields (have one dimensional scalar values for temperature at every point in space), Electromagnetic fields that are considered as vector fields (have three dimensional vector values at every point in space), and Gravitational fields that are considered as tensor fields (have four dimensional space-time tensor values at every point in space).
[37] Maxwell's Equations of Electromagnetism are as follows:

$$\nabla . E = \frac{\rho}{\varepsilon_0}$$

experimental measurement, c has been proven to be genuinely constant. It is assumed that c is one of the most fundamental constants of nature, and it is a significant number that shows up in many equations in physics. We have assumed that it is constant, and our current experiments have confirmed this assumption. Strangely enough, the speed of the electromagnetic field in vacuum or c is inversely related to the magnetic constant (permeability) or $\mu 0$ and the electric constant (permittivity) or $\varepsilon 0$, as follows:

$$c = \frac{1}{\sqrt{\mu_0 \varepsilon_0}}$$

This relation is derived from Maxwell's equations of electromagnetism, and it is how they have tested the speed of light and the accuracy of Maxwell's equations. We will use this equation later for comparative analysis of the constants in physics.

---

known as the Gauss's law, which states that electric charge creates electric force.

$\nabla . B = 0$

Known as the Gauss's law for magnetism, which states that magnetic forces are constant on the surface of any sphere around a magnetic source (or magnet).

$\nabla \times E = \frac{\partial B}{\partial t}$

Known as the Faraday's law, which states that moving, changing or fluctuating magnetic forces in time create electric forces.

$\nabla \times B = \mu_0 j + \mu_0 \varepsilon_0 \frac{\partial E}{\partial t} = \mu_0 j + \frac{1}{c^2} \frac{\partial E}{\partial t}$

Known as the Ampere's law, which states changing or fluctuating electric force in time creates magnetic force.

- **Fine Structure Constant (Dimensionless)**

$$\alpha = 0.0072973525693 \ldots$$

This dimensionless constant is related to the strength of the electromagnetic interaction between elementary charged particles. This constant is also known as the Somerfield's constant. The calculated value for α is:

$$0.0072973525693 \ldots \text{ or } \frac{1}{137.035999084 \ldots} \approx \frac{1}{137}$$

Is α truly constant, or is it constant at the levels of human measurements and experimental abilities? Observations have been made that suggest α has increased very slightly throughout the life of our universe. For all practical purposes, it is assumed that α, if not the most, but is one of the most fundamental dimensionless constants in physics. If α changes by a percent, the world, as we see, will not exist anymore.

Interesting set of mathematical relationships[38] exist between the fine structure constant α, magnetic constant $\mu 0$, electric constant $\varepsilon 0$, Plank constant h, elementary charge e, Coulomb constant k, the speed of electromagnetic field or wave (light) in vacuum c, and the mathematical transcendental number π. These are very unusual mathematical relationships between eight very important constants in mathematics and physics, which we will refer to throughout the book. These relationships are:

$$\alpha = \frac{e^2}{2\varepsilon_0 ch} = \frac{\mu_0 c e^2}{2h} = \frac{2\pi k e^2}{hc}$$

There have been various interpretations of the physical meaning

---

[38] For further explanation see the following site in Wikipedia.
https://en.wikipedia.org/wiki/Fine-structure_constant

of the fine structure constant α, and there are engaging discussions on the importance of this constant, and how it relates to the Fine-Tuning Principle[39].

- **Elementary Charge**

$$e = 1.6021766208 \times 10^{-19} \, (Coulomb)$$

The elementary charge, denoted by e or Q, has been defined as the electric charge carried by a single electron defined as being negative or, equivalently, a single proton defined as being positive. The elementary charge is one of the most fundamental constants in physics. There are elementary particles, such as quarks, that are known to have fractions (1/3) of elementary charges. Therefore, one can say that the "quantum of charge" is 1/3 e. We can, therefore, say that the "elementary charge" is three times as large as the quark charges. The calculated value for e is:

$$1.6021766208 \times 10^{-19} \, (Coulomb)$$

Again we can ask; is *e* truly constant, or is it constant at the levels of human measurements and experimental abilities? We assumed that it is truly constant unless proven otherwise. Again, a fascinating mathematical relationship exists between the elementary charge e, magnetic constant $\mu 0$, the electric constant $\varepsilon 0$, Plank constant h, the fine structure constant α, the speed of electromagnetic field or wave (light) in vacuum c and the irrational mathematical number $\sqrt{2}$. We see a very unusual

---

[39] The Fine Tuning Principle, states that all structures, values or ratios of important physical and mathematical constants (such as the fine structure constant α, mass ratio of proton to electron, gravitational constant, plank constant, expansion rate or mass density of the universe, speed of light, the value of π, e, and golden ratio to name a few) have been such finely tuned, without which the world as we know would have never been stable and existed. It is usually deduced that for the world to be and exist as it is, it had to be finely tuned and designed and caused by a grand designer.

mathematical relationship between seven crucial constants in mathematics and physics. They are shown below.

$$e^2 = \frac{2h\alpha}{\mu_0 c} = 2h\alpha\varepsilon_0 c$$

Or;

$$e = \sqrt{\frac{2h\alpha}{\mu_0 c}} = \sqrt{2h\alpha\varepsilon_0 c}$$

Or;

$$e = \sqrt{2}\sqrt{\frac{h\alpha}{\mu_0 c}} = \sqrt{2}\sqrt{h\alpha\varepsilon_0 c}$$

In here we have the value of the elementary charge of an electron or proton related to the square root of two (an irrational number), magnetic constant μ_0 and electric constant ε_0, Plank constant h, fine structure constant α, and the speed of electromagnetic field or wave (light) in vacuum c. We will later use these equations in a comparative analysis of various constants in physics.

- **Electron Rest Mass**

$$m_e = 9.1093 \ldots \times 10^{-31} \; (Kilograms)$$

The Electron rest mass me, is a fundamental constant number in physics. The rest mass[40] of an electron, which is a fundamental

---

[40] The need to specify rest or stationary mass is due to the fact that according to special relativity, if an object's speed is increased, its relativistic mass will also increase. The relativistic mass m^r, and the rest mass m (in here an electron m_e) of an object, are related through the following special relativistic expression;

$$m^r = \frac{m}{\sqrt{1-(\frac{v}{c})^2}} = \gamma m$$

Where v, is the actual speed of an object with the rest mass m, and γ is the Lorentz particle with an elementary charge *e* (negative sign) has a constant value equal to;

$$9.1093837015 \ldots \times 10^{-31} \, (Kilograms)$$

The Electron rest mass can be calculated through an exciting equation that relates it to the Plank constant h, the square of the fine structure constant α, the speed of electromagnetic field or wave (light) in vacuum c, and the Rydberg constant[41]. The equation is as follows;

$$m_e = \frac{2R_\infty h}{c\alpha^2}$$

---

factor which is equal to;

$$\gamma = \frac{1}{\sqrt{1 - (\frac{v}{c})^2}}$$

It is easy to see if an object does not move or $v = 0$, then $\gamma = 1$, and the relativistic mass equals to the rest mass or $m^r = m$. For an electron with a rest mass m_e, we get;

$$m_e^r = \frac{m_e}{\sqrt{1 - (\frac{v}{c})^2}} = \gamma m_e$$

And for a proton with a rest mass m_p, we get;

$$m_p^r = \frac{m_p}{\sqrt{1 - (\frac{v}{c})^2}} = \gamma m_p$$

---

[41] Rydberg constant $R\infty$, which we will cover later, is a constant in the electromagnetic spectra of an atom. It is equal to $1.0973731568160 \ldots \times 10^7$ (meter^{-1}). It is related to important constants in physics through the following equation;

$$R_\infty = \frac{m_e e^4}{8c\varepsilon_0^2 h^3}$$

This equation will become important in our analysis later. We want to know if *me* is truly constant, or is it constant at the levels of human measurements and experimental capabilities? We have assumed that on a rest mass basis, this is truly a constant physical number unless proven otherwise.

- **Proton-Electron Mass Ratio (Dimensionless)**

$$\mu = \frac{m_p}{m_e} = 1836.15267343\ldots$$

The proton-electron rest mass ratio μ is a fundamental constant dimensionless number in physics. It is the ratio of the rest mass of a proton to the rest mass of an electron. This number has been calculated to a certainty of 0.1 parts per billion. Proton-Electron mass ratio μ and the fine structure constant α are the two primary and critical dimensionless quantities in physics. The calculated value for μ is:

$$1836.15267343\ldots or \approx 1836$$

We have to ask the same questions as before. Is μ constant in all space-time situations such as highly curved spaces near black holes or at the time of big bang? Is μ constant through time? Is μ constant in other possible universes? Is μ truly constant, or is it constant at the levels of human measurements and experimental abilities? Astrophysicists have observed a minimal change in the value of this ratio through the history of our universe. For all practical purposes, it is assumed that μ is one of the most fundamental dimensionless constants in physics.

- **Rydberg Constant**

$$R_\infty = 1.0973\ldots \times 10^7 \ (meters^{-1})$$

Rydberg constant $R\infty$ is a constant number related to the electromagnetic spectra of an atom. This number represents the limiting value of the largest wave number or the inverse of the wavelength of any photon that can be detected from an atom or ion. The value for this constant is equal to;

$$1.0973731568160 \dots \times 10^7 \ (meter^{-1})$$

Or;

$$13.605693009 \dots \ eV$$

It is also related to essential constants in physics through the following equation;

$$R_\infty = \frac{m_e e^4}{8c\varepsilon_0^2 h^3} = \frac{\alpha^2 m_e c}{2h}$$

The above equation is used in approximate calculations related to the spectrum of hydrogen and other atoms and ions.

- **Boltzmann Constant**

$$k_b = 1.3806 \dots \times 10^{-23} \ (Joules.Kelvin^{-1})$$

This constant is named after Ludwig Eduard Boltzmann, the 19th century famous Austrian physicist and philosopher. It is a dimensional constant number (in *Joules J*, per *Kelvin temperature K*), that shows the equivalency relationship of the average kinetic energy of gas particles (a microscopic parameter) to the gas temperature (a macroscopic parameter). It is a constant proportionality factor that relates gas temperature to its kinetic energy. This constant shows up in different areas of physics, such as in the gas constant parameter, Plank's black body law, and the equation for entropy (the second law of thermodynamics). The value for this constant is equal to;

$$1.38064852 \dots \times 10^{-23} \ (Joules.Kelvin^{-1})$$

Two relevant equations incorporating Boltzmann constant are;

- ✓ Ideal gas Law

$$PV = Nk_bT$$

Or;

$$K_b = \frac{PV}{NT}$$

Where P is the gas pressure, V is the gas Volume, N is the number of molecules in a gas, and T is the temperature.

- ✓ Entropy or second law of thermodynamics

$$S = k_b \ln W$$

Or;

$$K_b = \frac{S}{\ln W}$$

Where S is the Entropy or the degree of disorder and randomness of an isolated system in equilibrium (a macroscopic parameter), $\ln W$ is the natural logarithm of the number of possible microscopic states W of the gas. We have a statistical mechanic interpretation of the second law of thermodynamics or entropy. It can be seen that the dimension for Boltzmann constant is similar to the dimension of Entropy, in Joules per Kelvin.

- ○ **Stefan-Boltzmann Constant**
$\sigma = 5.6703 \ldots \times 10^{-8} \; (Watts.Meters^{-2}.Kelvin^{-4})$

This constant dimensional number is named jointly after Ludwig Eduard Boltzmann and Josef Stefan, 19th century Austrian physicists, mathematicians, and philosophers who pioneered the theory of thermal radiation. The Stefan-Boltzmann constant is a significant dimensional number (in units of *Watts W*, per square *Meters M*, per fourth power of *Kelvin temperature K*), which shows the proportionality relationship of the total intensity of all radiations emitted from a black body[42] to its temperature. This constant can be used to measure the amount of heat created and emitted by a black body. The calculated value for this constant is equal to;

$$5.670374419\ldots \times 10^{-8} \ (Watts.Meters^{-2}.Kelvin^{-4})$$

The Stefan-Boltzmann constant is related to other constants of physics and mathematics through the following equations;

$$\sigma = \frac{2\pi^5 k_b^4}{15c^2 h^3} = \frac{Total\ Flux_{Black\ Body}}{T^4}$$

It is through these relationships that Astrophysicists can calculate the effective temperatures at the surface of faraway stars. It is incredible to see such a constant at play for understanding the temperature behavior of distant galaxies. At the same time, it can be calculated through the transcendental constant number π, Boltzmann constant Kb, speed of electromagnetic field or wave in vacuum c, and Plank constant h.

- **Standard Model of Particle Physics Constants**
  Fermions, Bosons, Coupling Constants, etc.

  To understand constants used in the Standard Model of Particle Physics[43] or SMPP, we need to briefly understand what the

---

[42] Black Body is a physical body absorbing all electromagnetic radiations. Ideally it will absorb radiation from all frequencies, can have temperature, and emit radiation back to the outside. For further description see, https://en.wikipedia.org/wiki/Black_body.
[43] For further information on the Standard Model of Particle Physics see; https://en.wikipedia.org/wiki/Standard_Model

Standard model is. SMPP is the most recent model for the dynamics of elementary particles in the universe. It has been relatively accurate in its predictions confirmed by experiments, and it is a consolidated model of the Electromagnetic, Strong and Weak forces[44]. It does not include the Gravitational force as a part of the model. To understand all the existing forces and particles of nature, we rely on the SMPP and the Gravitational Theory. Based on the SMPP theory there are **12 Fermions** or elementary *matter particles* (all with ½ particle spin) which are divided into two distinct groups, **6 Quarks** (called **u**p, **d**own, **c**harm, **s**trange, **t**op, **b**ottom), and **6 Leptons** (called **e**lectron, **e**lectron **n**eutrino, **m**uon, **m**uon **n**eutrino, **t**au, **t**au **n**eutrino). In connection with these matter particles, and how they interact with one another,

---

[44] There are four known fundamental forces in nature. The Gravitational, Electromagnetic, Strong and Weak forces. Each has different effective ranges. Gravitation and Electromagnetic forces have infinite effective ranges. The Strong force has an effective range of 10^{-15} meter and the Weak force at 10^{-18} meter or at the nuclear and Quantum Mechanical ranges. The relative strength of these forces in relation to the Gravitational force is very different. If we assume the strength of the Gravitational force to be 1, then the Strong force will have 10^{+38} times strenght, Electromagnetic force 10^{+36} times strength, and the weak force 10^{+25} times strength. The Strong and Weak forces are relatively very strong at very small scales and the Electromagnetic force is relatively very strong at all distances. Gravitational force is only strong at large distances. According to the Standard Model of Particle Physics the force carrying particles for each force is different. The force carrying particle for the Gravitational force is not yet known but it is assumed to be Gravitons (based on the theory of General Relativity, GR), for Electromagnetic force is Photons (based on the Quantum Electrodynamics Theory, QED), for the Strong force is Gluons (based on the Quantum Chromo dynamics Theory, QCD), and for the Weak force is W and Z Bosons (based on the Electroweak Theory, EWT). The General Theory of Relativility in tensor forms is written as:

$$\frac{8\pi G}{C^4} T_{\mu\nu} = R_{\mu\nu} - \frac{1}{2} g_{\mu\nu} R$$

And the Standard Model of Particle Physics in tensor form is written as:

$$\mathcal{L} = -\frac{1}{4} F_{\mu\nu} F^{\mu\nu} + i\bar{\Psi} D\Psi + D_\mu \Phi^\dagger D^\mu \Phi + \bar{\Psi}_L \bar{Y} \Phi Y_R - V(\Phi)$$

there are **6 Bosons** or elementary *force particles* or force-carrying particles (called **photons** causing electromagnetic force interactions, **gluons** causing strong nuclear force interactions, **W+, W−,** and **Z** bosons causing weak nuclear force interactions, and **Higgs** bosons causing the particle mass effects). The SMPP model breaks down the elementary particle universe into 12 Fermions as matter particles, and 6 Bosons as force particles (particles that cause and carry the fundamental Electromagnetic, Strong and Weak forces). There is a wonderful and relatively complicated mathematical relationship that describes the general dynamics and interaction of the matter particles through the force particles. Through the general equation of the Standard model, we are required to have several physical constants or inputs to allow the utilization of the model. They are:

- ✓ **Twelve Fermion mass**[45] **constants** which include:
    - Six **Quark** masses

        $$\text{Up} \approx 0.0022 \; \frac{\text{GeV}}{c^2}$$
        $$\text{Down} \approx 0.0047 \; \frac{\text{GeV}}{c^2}$$
        $$\text{Charm} \approx 1.2800 \; \frac{\text{GeV}}{c^2}$$
        $$\text{Strange} \approx 0.0960 \; \frac{\text{GeV}}{c^2}$$
        $$\text{Top} \approx 173.21 \; \frac{\text{GeV}}{c^2}$$
        $$\text{Bottom} \approx 4.1800 \; \frac{\text{GeV}}{c^2}$$

    - Six **Lepton** masses

        $$\text{electron} \approx 0.5110 \; \frac{\text{MeV}}{c^2}$$
        $$\text{electron neutrino} \approx 2.2000 \; \frac{\text{eV}}{c^2}$$
        $$\text{muon} \approx 105.66 \; \frac{\text{MeV}}{c^2}$$
        $$\text{muon neutrino} \approx 0.1700 \; \frac{\text{MeV}}{c^2}$$

---

[45] To change particle mass from GeV/c^2 to Kg you have:

$$1 \; \text{GeV}/c^2 = 1.7827 \times 10^{-27} \; \text{Kg}$$

$$\text{tau} \approx 1.7768 \, \frac{\text{GeV}}{c^2}$$
$$\text{tau neutrino} \approx 18.200 \, \frac{\text{MeV}}{c^2}$$

- ✓ **Six Boson mass constants** which include:

$$\text{photon} \approx 0.000 \, \frac{\text{GeV}}{c^2}$$
$$\text{gluon} \approx 0.000 \, \frac{\text{GeV}}{c^2}$$
$$W+ \approx 80.379 \, \frac{\text{GeV}}{c^2}$$
$$W- \approx 80.379 \, \frac{\text{GeV}}{c^2}$$
$$Z \approx 91.187 \, \frac{\text{GeV}}{c^2}$$
$$\text{Higgs} \approx 124.97 \, \frac{\text{GeV}}{c^2}$$

- ✓ **Three force coupling dimensionless constants**:
  - Electromagnetic force constant (Fine structure) $\quad \alpha \approx \frac{1}{137}$
  - Strong force constant $\quad \alpha_s \approx 1$
  - Weak force constant $\quad \alpha_w \approx 3 \times 10^{-7}$

- ✓ **Four other constants**:
  - One CP-violating phase, which we will not analyze here
  - Three CKM mixing angles, which we will not analyze here

The Standard model requires 25 constants of nature. We have already analyzed and considered two of them before, the mass of the electron e, and the fine structure constant α. Going forward, excluding the zero masses of photons and gluons, we *will consider an additional 21 constants of physics*. They include; particle mass for 6 Quarks and five remaining Leptons (excluding electron mass), in addition to 4 Bosons masses (excluding photons and gluons) and 2 Strong and Weak coupling force constants (excluding the fine structure constant α), plus 3 CKM mixing angles constants, and 1 CP-violating phase constant. We could make the 18 Fermion and Boson constant masses dimensionless by calculating their relative masses in terms of Higgs mass (how many multiple masses are required to make 1

Higgs mass). The result gives us dimensionless relative mass constants for all the 18 Fermions and Bosons. These results are shown as follows:

- **Twelve dimensionless Fermion mass constants** (multiples required to make one Higgs mass):
    - Six **Quarks**
        - Up   ≈ 56805
        - Down   ≈ 26589
        - Charm   ≈ 97.632
        - Strange   ≈ 1301.8
        - Top   ≈ 0.7215
        - Bottom   ≈ 29.897
    - Six **Leptons**
        - electron   ≈ 244560
        - electron neutrino ≈ 5.68×10^{13}
        - muon   ≈ 1182.8
        - muon neutrino   ≈ 735118
        - tau   ≈ 70.334
        - tau neutrino   ≈ 6866.5

- **Six dimensionless Boson mass constants** (multiples required to make one Higgs mass):
    - photon   ≈ 0.000
    - gluon   ≈ 0.000
    - W+ boson   ≈ 1.5548
    - W- boson   ≈ 1.5548
    - Z boson   ≈ 1.3705
    - **Higgs** boson   = 1.0000

Among the 12 Fermions and 6 Bosons, the largest particle mass belongs to the Top Quark and the smallest to photon and gluon. Among the 6 Quarks, the smallest masses belong to the Up and Down Quarks. Generally speaking, all force-carrying particles or Bosons, except photons and gluons, are heavy and massive particles. Less heavy particles such as Up and Down Quarks, electrons, and all three types of neutrinos are more stable and lasting. More massive particles are more unstable, difficult to observe, and do not last long. We must remember that the 18 Fermion and Boson particles have additional features such as spin (which can be ½, 0, 1) and charge (which can vary from -1, -2/3, -1/3, 0, 1/3, 2/3, 1) that are fingerprints associated with them. These additional fundamental properties or constants could also be incorporated in the list of our constant numbers in physics. For now, we will only focus on 21 constants of the Standard model, as described so far.

The existence of structures and mathematical relationships between constant numbers that play critical roles in the growth patterns, dynamic, and physical behavior in nature is fascinating. Even more impressive is the meaning of the existing mathematical relationships between them.

A summary of the constant numbers in physics that we have covered so far is shown below:

Summary of the reviewed constant numbers in physics

Constant	Sign	≈ Calculated Value	Units
Gravitational	G	$6.6740 \ldots \times 10^{-11}$	$(M^3.Kg^{-1}.S^{-2})$
Magnetic	μ_0	$1.2566 \ldots \times 10^{-6}$	$(H.M^{-1})$
Electric	ε_0	$8.8541 \ldots \times 10^{-12}$	$(F.M^{-1})$
Plank	h	$6.6260 \ldots \times 10^{-34}$	$(J.S)$
Speed of Light	c	2.99792458×10^8	$(M.S^{-1})$
Fine Structure	α	$7.2973 \ldots \times 10^{-3}$	None
Elementary Charge	e	$1.6021 \ldots \times 10^{-19}$	(C)
Electron Rest Mass	m_e	$9.1093 \ldots \times 10^{-31}$	(Kg)
$m_{proton}/m_{electron}$	μ	1836.1526	None
Rydberg	R_∞	$1.0973 \ldots \times 10^7$	(M^{-1})
Boltzmann	K_b	$1.3806 \ldots \times 10^{-23}$	$(J.K^{-1})$
Stefan-Boltzmann	σ	$5.6703 \ldots \times 10^{-8}$)	$(W.M^{-2}.K^{-4})$

The 21 new constant numbers from the Standard Model of Particle Physics or SMPP are summarized below:

Summary of the reviewed constant numbers from the SMPP

Constant	Sign	≈ Calculated Value	Units
Up Quark Mass	m_{uq}	0.0022	$(Gev.C^{-2})$
Down Quark Mass	m_{dq}	0.0047	$(Gev.C^{-2})$
Charm Quark Mass	m_{cq}	1.2800	$(Gev.C^{-2})$
Strange Quark Mass	m_{sq}	0.0960	$(Gev.C^{-2})$
Top Quark Mass	m_{tq}	173.21	$(Gev.C^{-2})$
Bottom Quark Mass	m_{bq}	4.1800	$(Gev.C^{-2})$
E Neutrino Mass	m_{en}	2.2000	$(ev.C^{-2})$
Muon Mass	m_m	105.66	$(Mev.C^{-2})$
M Neutrino Mass	m_{mn}	0.1700	$(Mev.C^{-2})$
Tau Mass	m_t	1.7768	$(Gev.C^{-2})$
T Neutrino Mass	m_{tn}	18.200	$(Mev.C^{-2})$
W+ Boson Mass	m_{w+}	80.379	$(Gev.C^{-2})$
W- Boson Mass	m_{w-}	80.379	$(Gev.C^{-2})$
Z Boson Mass	m_z	2.2000	$(Gev.C^{-2})$
Higgs Mass	m_h	124.97	$(Gev.C^{-2})$
Strong Force	α_s	1	None
Weak Force	α_w	3×10^{-7}	None
CP Violating Phase	c_p	*Not analyzed*	-
CKM Mixing Angles	ckm_θ	*Not analyzed*	-
CKM Mixing Angles	ckm_θ	*Not analyzed*	-
CKM Mixing Angles	ckm_θ	*Not analyzed*	-

Interesting observations and mathematical relationships exist among our reviewed constant numbers in physics. They are:

- *The relationships and symmetries between the reviewed physical constants are fascinating. Some constants such as α, e, and σ are even related to the transcendental number π and the irrational number $\sqrt{2}$. These relationships indicate that there could be more fundamental interrelationships between the equations in physics and mathematics.*

- *The following interrelationships between the analyzed constants are noteworthy. The gravitational constant is the only physical constant that is not expressed in terms of the other ones.*

$$\mu_0 = \frac{2\alpha h}{e^2 c} = \frac{1}{\varepsilon_0 c^2}$$

$$\varepsilon_0 = \frac{1}{\mu_0 c^2} = \frac{1}{4\pi k} = \frac{e^2}{2\alpha hc}$$

$$h = \frac{8\alpha}{\mu_0 c K_j^2}$$

$$c = \frac{1}{\sqrt{\mu_0 \varepsilon_0}}$$

$$\alpha = \frac{e^2}{2\varepsilon_0 ch} = \frac{\mu_0 ce^2}{2h} = \frac{2\pi ke^2}{hc}$$

$$e = \sqrt{2}\sqrt{\frac{h\alpha}{\mu_0 c}} = \sqrt{2}\sqrt{h\alpha \varepsilon_0 c}$$

$$m_e = \frac{2R_\infty h}{c\alpha^2}$$

$$R_\infty = \frac{m_e e^4}{8c\varepsilon_0^2 h^3} = \frac{\alpha^2 m_e c}{2h}$$

$$\sigma = \frac{2\pi^5 k_b^4}{15 c^2 h^3}$$

These mathematical relationships indicate the probability of more intricate underlying physical laws and mathematical structures that we are not aware of. The following matrix shows the existing mutual interrelationships (+ sign) among the important constant in physics (except the SMPP model), analyzed so far.

	G	μ_0	ε_0	h	c	α	e	m_e	R_∞	σ
G										
μ_0			+	+	+	+	+			
ε_0		+		+	+	+	+		+	
h		+	+		+	+	+	+	+	+
c		+	+	+		+	+	+	+	+
α		+	+	+	+		+	+	+	
e		+	+	+	+	+			+	
m_e				+	+	+			+	
R_∞			+	+	+	+	+	+		
σ				+	+					
Total	0	5	6	8	8	7	6	4	6	2

As seen from the above table, h, c, α, e, μ_0, ε_0, and R_∞ are highly interrelated. They are constants related to the Electromagnetic force, which may suggest an underlying unknown structure that, if found, could consolidate some of these constants. We also know that based on the Standard Model, the Fine structure constant α (dimensionless) seems to be a fundamental number that shows up in most dynamic situations.

Not including the Standard model, there are above thirty important constants in physics. So far, we have reviewed 12 of the most important ones plus 21 from the Standard model. We did not review the Cosmological constant[46] due to the uncertainties surrounding its meanings and interpretation.

We can always create new constants from the existing ones. The ratio of charge to mass or charge density for various fundamental particles such as protons, all six quarks, and Leptons (which includes electrons) can be considered as new constants. For electrons the charge mass ratio is $1.7588... \times 10^{11}$ Coulombs per Kilograms, and for protons it is $9.5789... \times 10^{7}$ Coulombs per Kilograms. Interestingly enough, electrons have approximately 1836 times more charge density compared to protons or equal to the ratio of the two masses. Other interesting constants can be shown as the ratio of Gravitational potential of two fundamental particles such as two protons or two electrons to their rest energies, which can be defined as their Gravitational coupling constants. Such constants have been calculated, and for electrons, it is to the order of 10^{-45}, and for protons to the order of 10^{-39}. Another interesting dimensionless constant is the ratio of the gravitational force between two electrons at the de Broglie wavelength to the electric force between these two electrons at the same wavelength. We can call it the constant of Omega, and it is calculated as follows:

$$\Omega = \frac{\frac{Gm_e^2}{hc}}{\frac{e^2}{4\pi\varepsilon_0 hc}} = 4\pi\varepsilon_0 G \left(\frac{m_e}{e}\right)^2 = 4.2222... \times 10^{-32}$$

Even though we can measure these constants with reasonable accuracy, we do not know the reasons for them having their measured values. Einstein hoped that these physical constants could be explained by the Unified Theory of Everything. For now, we try to measure, use, understand, and analyze them to the best of our abilities.

---

[46] Cosmological constant is defined as the energy density of space, or energy in vacuum. It is related to the concept of dark energy. It is still a controversial concept. For more information see: https://en.wikipedia.org/wiki/Cosmological_constant

We have assumed that the values for these constants are resilient, time-invariant, stable, and non-changing. Some research has been done on the stability of the Fine structure constant through time and shows that it has changes minutely through time. We still need to research the nature of the physical constants and the possibility of unifying them. As Einstein had hoped:

> *"Dimensionless constants in the laws of nature, which from the purely logical point of view can just as well have different values, should not exist or be reduced to geometrical constants π, e, φ, etc.."*

In other words, the mathematical expressions for the final unifying equation of nature should only have geometrical constants π, e, φ, etc.

Related Quotes

1. "If the deep logic of what determines the value of the fine-structure constant also played a significant role in our understanding of all the physical processes in which the fine-structure constant enters, then we would be stymied. Fortunately, we do not need to know everything before we can know something."

   John D. Barrow, New Theories of Everything
   https://www.goodreads.com/quotes/tag/constants-of-nature

2. "We can measure the fine structure constant with very great precision, but so far, none of our theories has explained its measured value. One of the aims of the superstring theory is to predict this quantity precisely. Any theory that could do that would be taken very seriously indeed as a potential Theory of Everything.'"

   John D. Barrow, Impossibility: The Limits of Science and the Science of Limits
   https://www.goodreads.com/quotes/tag/constants-of-nature

3. "Since only a narrow range of the allowed values for, say, the fine structure constant will permit observers to exist in the universe, we must find ourselves in the narrow range of possibilities which permit them, no matter how improbable they are. We must ask for the conditional probability of observing constants to take particular ranges, given that other features of the Universe, like its age, satisfy necessary conditions for life."

   John D. Barrow, The Constants of Nature: The Numbers That Encode the Deepest Secrets of the Universe
   https://www.goodreads.com/quotes/tag/constants-of-nature

4. "Considerable mysteries are surrounding the strange values that

Nature's actual particles have for their mass and charge. For example, there is the unexplained 'fine structure constant' ... governing the strength of electromagnetic interactions, ..."

Roger Penrose, The Road to Reality: A Complete Guide to the Laws of the Universe
https://www.goodreads.com/quotes/tag/constants-of-nature

5. "Nature is so constituted that it is possible logically to lay down such strongly determined laws that within these laws only rationally determined constants occur (not constants, therefore, whose numerical value could be changed without destroying the theory)."

Albert Einstein, Autobiographical Notes
https://www.goodreads.com/quotes/tag/constants-of-nature

6. "... it should be remembered that the atomicity of electric charge has already found its expression in the specific numerical value of the fine structure constant, a theoretical understanding of which is still missing today."

Wolfgang Pauli, Theory of Relativity
https://www.goodreads.com/quotes/tag/constants-of-nature

Questions

1. **What is the energy equivalent equation for the rest mass of an electron in terms of h, a, $\mu 0$, $\varepsilon 0$, and $R\infty$?**

   Solution:

   $$E_e = m_e c^2$$

   And;

   $$m_e = \frac{2R_\infty h}{c\alpha^2}$$

   And;

   $$c = \frac{1}{\sqrt{\mu_0 \varepsilon_0}}$$

   Therefore;

   $$E_e = \frac{2R_\infty h}{\sqrt{\mu_0 \varepsilon_0}\, \alpha^2}$$

2. What is the energy of proton with frequency v in terms of α, μ0, ε0, v, and Kj?

   Solution:

   $$E_p = hv$$

   And;

   $$h = \frac{8\alpha}{\mu_0 c K_j^2}$$

   And;

   $$c = \frac{1}{\sqrt{\mu_0 \varepsilon_0}}$$

   Therefore;

   $$E_e = \frac{8\alpha v \sqrt{\mu_0 \varepsilon_0}}{\mu_0 K_j^2}$$

3. Find the Electric charge constant in terms of h, α, μ0, ε0, and the silver ratio φ2.

   Solution:

   $$e = \sqrt{2}\sqrt{\frac{h\alpha}{\mu_0 c}} = \sqrt{2}\sqrt{h\alpha\varepsilon_0 c}$$

   And;

   $$\sqrt{2} = \varphi_2 - 1$$

   Therefore;

$$e = (\varphi_2 - 1)\sqrt{\frac{h\alpha\sqrt{\mu_0\varepsilon_0}}{\mu_0}} = (\varphi_2 - 1)\sqrt{h\alpha\varepsilon_0 \frac{1}{\sqrt{\mu_0\varepsilon_0}}}$$

4. **Find the Plank density for a Plank cube with sides equal to Plank length and mass equal to Plank mass.**

Solution:

$$\text{Plank Length} = \left(\frac{HG}{c^3}\right)^{\frac{1}{2}} = 1.6162\ldots \times 10^{-35} \text{ m}$$

$$\text{Plank Cube Volume} = \left(\frac{HG}{c^3}\right)^{\frac{3}{2}} = 4.2219\ldots \times 10^{-105} \text{ m}^3$$

$$\text{Plank Mass} = \left(\frac{HC}{G}\right)^{\frac{1}{2}} = 2.1764\ldots \times 10^{-8} \text{ kg}$$

Therefore;

$$\text{Plank Cube Density} = \frac{\text{Plank Mass}}{\text{Plank Volume}} = \frac{2\pi c^5}{hG^2} = 5.1552\ldots \times 10^{96} \frac{kg}{m^3}$$

5. **Find the Plank density for a Plank Cube with sides equal to Plank length and mass equal to Plank mass, in terms of h, G, μ_0, ε_0.**

Solution:

We know;

$$c = \frac{1}{\sqrt{\mu_0\varepsilon_0}}$$

$$\text{Plank Cube Density} = \frac{2\pi c^5}{hG^2} = \frac{2\pi\left(\frac{1}{\sqrt{\mu_0\varepsilon_0}}\right)^5}{hG^2}$$

$$\text{Plank Cube Density} = \frac{2\pi}{hG^2(\mu_0\varepsilon_0)^{\frac{5}{2}}} = 5.1552\ldots \times 10^{96} \frac{kg}{m^3}$$

Incredibly, π, h, G, μ_0, ε_0 can determine Plank Cube density.

6. **Find the Plank density for a Plank Sphere with a diameter equal to Plank length and mass equal to Plank mass.**

   Solution:

   $$Plank\ Lenght = \left(\frac{HG}{C^3}\right)^{\frac{1}{2}} = 1.6162\ldots \times 10^{-35}\ m$$

   $$Plank\ Sphere\ Volume = \frac{\pi}{6} \left(\frac{HG}{C^3}\right)^{\frac{3}{2}} = 2.2105\ldots \times 10^{-105}\ m^3$$

   $$Plank\ Mass = \left(\frac{HC}{G}\right)^{\frac{1}{2}} = 2.1764\ldots \times 10^{-8}\ kg$$

   Therefore;

   $$Plank\ Sphere\ Density = \frac{Plank\ Mass}{Plank\ Volume} = \frac{6C^5}{\pi HG^2} = \frac{12C^5}{hG^2}$$

   $$Plank\ Sphere\ Density = 9.8453\ldots \times 10^{96}\ \frac{kg}{m^3}$$

7. **Find the Plank density for a Plank Sphere with a diameter equal to Plank length and mass equal to Plank mass, in terms of h, G, μ_0, ε_0.**

   Solution:

   We know;

   $$c = \frac{1}{\sqrt{\mu_0 \varepsilon_0}}$$

   $$Plank\ Sphere\ Density = \frac{2\pi C^5}{hG^2} = \frac{12}{hG^2(\mu_0\varepsilon_0)^{\frac{5}{2}}}$$

   $$Plank\ Sphere\ Density = 9.8453\ldots \times 10^{96}\ \frac{kg}{m^3}$$

   Interestingly enough, Plank Sphere density is not a function of π, but Plank Cube density is. The ratio of Plank Sphere density to Plank Cube density is equal to $6/\pi$ or ≈ 1.9098. Plank Sphere because of lower volume ($\pi/6 \approx 0.5235$ or almost half of the volume of a Plank cube) must have 1.9098 times higher density to capture the same Plank mass.

8. The rest mass of a proton is $1.672621898 \times 10^{-27}$ kg, and of an electron is $9.10938356 \times 10^{-31}$ kg. Determine their masses in terms of multiples of Plank mass.

Solution:

$$\text{Proton rest mass} = 1.672621898 \times 10^{-27} \text{ kg}$$

$$\text{Electron rest mass} = 9.10938356 \times 10^{-31} \text{ kg}$$

$$\text{Plank mass} = 2.1764\ldots \times 10^{-8} \text{ kg}$$

Therefore;

$$\frac{\text{Proton rest mass}}{\text{Plank mass}} = 7.68527 \times 10^{-20}$$

$$\frac{\text{Electron rest mass}}{\text{Plank mass}} = 4.18553 \times 10^{-23}$$

The rest mass of a Proton and an Electron is smaller than the critical upper limit of Plank mass. If their masses were equal or larger than the Plank mass, they would probably have turned into tiny black holes.

9. If the volume of the observable universe is 4×10^{80} m³, determine how many Plank Cubes or Spheres can exist in this volume.

Solution:

$$\text{Plank Cube Volume} = \left(\frac{HG}{c^3}\right)^{\frac{3}{2}} = 4.2219\ldots \times 10^{-105} \text{ m}^3$$

$$\text{Plank Sphere Volume} = \frac{\pi}{6}\left(\frac{HG}{c^3}\right)^{\frac{3}{2}} = 2.2105\ldots \times 10^{-105} \text{ m}^3$$

The volume of the Universe $\qquad 4 \times 10^{80}$ m³

Therefore;

$$\text{Plank Cubes in the Universe} = \frac{Volume\ of\ the\ Univesre}{Volume\ of\ Plank\ Cube}$$

$$\text{Plank Cubes in the Universe} = 9.47441\ldots \times 10^{184}$$

$$\text{Plank Spheres in the Universe} = \frac{Volume\ of\ the\ Univesre}{Volume\ of\ Plank\ Sphere}$$

$$\text{Plank Spheres in the Universe} = 1.80955\ldots \times 10^{185}$$

If the universe was made up of tiny Plank spheres, there would be 1.9098 (or 6/π more) more spheres as compared to tiny Plank cubes.

10. **Determine the rest mass ratio of six quarks as multiples of the electron rest mass.**

    Solution:

Electron rest mass	Ratio = 1.0000
Up Quark rest mass	Ratio = 4.3053
Down Quark rest mass	Ratio = 9.1977
Charm Quark rest mass	Ratio = 2504.9
Strange Quark rest mass	Ratio = 187.89
Top Quark rest mass	Ratio = 338964
Bottom quark rest mass	Ratio = 8180.1

    The higher relative ratio of a Quark rest mass leads to less possibility of observing them at lower energies. Top Quarks are only observed at very high energies and are very unstable.

Final Note

So far, we have described and shown the values for important fundamental mathematical and physical constants. The puzzling issue is that we do not have any understanding of why they have their specific values.

It seems as if the mathematical and physical constants carry codes that need to be decoded. Certainly, they play a vital role in the original design, equations, dynamics, and required initial conditions of our universe. These fundamental constants are independent of our existence and seem to have a meaning of their own.

In this book, we tried to explain the concept of numbers and analyze 11 important constant numbers in mathematics (not including the Metallic ratios). We then tried to explain and analyze 33 fundamental constant numbers in physics (including 21 numbers from the Standard Model of Particle Physics). In total, we have reviewed and analyzed 44 important fundamental constants in Mathematics and Physics.

We presented the following mathematical interrelationships among fundamental irrational mathematical constants and complex numbers:

$$\sqrt{2} = 2\,Cos\left(\frac{\pi}{4}\right) = 2\,Sin\left(\frac{\pi}{4}\right) = \varphi_2 - 1$$

$$\sqrt{5} = \frac{1}{Sin\left(\frac{\pi}{10}\right)} - 1 = 2\,\varphi_1 - 1$$

$$\varphi_1 = 2\cos\left(\frac{\pi}{5}\right) = e^{\frac{+i\pi}{5}} + e^{\frac{-i\pi}{5}} = \frac{1 \pm \sqrt{5}}{2}$$

$$\varphi_2 = (2\cos(\frac{\pi}{4})) + 1 = e^{\frac{+i\pi}{4}} + e^{\frac{-i\pi}{4}} + 1 = \sqrt{2} + 1$$

$$\pi = \lim_{n\to\infty} \frac{e^{2n} n!^2}{2n^{2n+1}} = \sqrt{6 \sum_{n=1}^{n=\infty} \frac{1}{n^2}} = \sqrt{6\zeta(2)}$$

$$\frac{2}{\pi} = \left(\frac{\sqrt{2}}{2}\right)\left(\frac{\sqrt{2+\sqrt{2}}}{2}\right)\left(\frac{\sqrt{2+\sqrt{2+\sqrt{2}}}}{2}\right)\ldots$$

$$e^{i\pi} = -1 = i^2$$

$$\zeta(3) = \frac{1}{2}\int_0^\infty \frac{x^2}{(e^x-1)}dx = \frac{7\pi^3}{180} - 2\sum_{n=1}^{n=\infty} \frac{1}{n^3.(e^{2\pi n}-1)}$$

$$\gamma = -\int_0^\infty e^{-x} \ln x\, dx = \sum_{n=1}^{n=\infty}\left(\frac{1}{n} - \ln\frac{n+1}{n}\right)$$

$$\delta = \pi + \tan^{-1}(e^\pi)$$

$$\sqrt[i]{i} = i^{-i} = e^{\frac{\pi}{2}}$$

We also showed the following interrelationships among the fundamental constants in physics:

$$\mu_0 = \frac{2\alpha h}{e^2 c} = \frac{1}{\varepsilon_0 c^2}$$

$$\varepsilon_0 = \frac{1}{\mu_0 c^2} = \frac{1}{4\pi k} = \frac{e^2}{2\alpha hc}$$

$$h = \frac{8\alpha}{\mu_0 c K_J^2}$$

$$c = \frac{1}{\sqrt{\mu_0 \varepsilon_0}}$$

$$\alpha = \frac{e^2}{2\varepsilon_0 ch} = \frac{\mu_0 ce^2}{2h} = \frac{2\pi ke^2}{hc}$$

$$e = \sqrt{2}\sqrt{\frac{h\alpha}{\mu_0 c}} = \sqrt{2}\sqrt{h\alpha\varepsilon_0 c}$$

$$m_e = \frac{2R_\infty h}{c\alpha^2}$$

$$R_\infty = \frac{m_e e^4}{8c\varepsilon_0^2 h^3} = \frac{\alpha^2 m_e c}{2h}$$

$$\sigma = \frac{2\pi^5 k_b^4}{15c^2 h^3}$$

After reviewing the amazing relationships between the 44 reviewed constants through the 20 interrelated equations, we are triggered to think that there are unknown underlying mathematical and physical structures that we should search for. Amazingly, such inexact constant numbers (including Irrationals and Transcendentals) end up having such surprising and precise mathematical and physical relationships and patterns.

The minimum that we can conclude from our analysis is that the dimensionless irrational and transcendental constant numbers π, e, and $\varphi 1$ show up in all harmonic, repetitive, circular, and growing dynamic behaviors in nature. Also, the physical constant numbers G, α, h, c, e show up in most physical laws and dynamic behaviors. Even more interesting is that there are interrelationships between the Mathematical and Physical constants. Therefore we could also conclude that:

"The important fundamental constants in Mathematics and Physics are π, e, φ1, G, α, h, c, e because they show up everywhere."

We should also provide an answer to the question of; is mathematics invented or discovered?

Are numbers independent from the mathematical or physical concepts that apply to them? Is number π an independent entity etched as a universal code for the behavior of any circular motions, or number e for any growth, and φ for the combination of these two behaviors? Is the speed of light c, Plank constant h, etc. separate from the equations that use them?

Mathematic is a highly developed abstract language used by humans to describe shapes, patterns, and dynamic behavior of the world, realities, and existence around us. Through time, it has evolved and will continue to evolve in terms of scope, the broadness of application, abstractness of concepts, and completeness. Modern Mathematics is based on a set of logical and abstract rules invented based on patterns discerned by our brains. Could other advanced beings create a different kind of mathematics and logical rules? Probably yes! But their mathematics will only be valid if they can explain the shapes, patterns and forecast dynamic behavior of this world as accurately, consistently, and effectively or even better.

For now, mathematics and its applications in physics that we are aware of, are the most evolved and accurate and has helped us understand the universe around us effectively. The world that mathematics and physics try to explain is based on numbers (real or imaginary), geometrical shapes (real or abstract), and physical objects with some mathematical relationships that govern their behavior and dynamics. Numbers are an abstract invention of our mind, but they represent a feature in the universe that is not an invention. Through numbers and their mathematical relationships, we discover the reality of that feature that the numbers are representing. These features are real and are out there.

Two stars having gravitational effects on one another are always two interacting stars regardless of our existence. Aliens on other planets, might name these two stars differently, utilize their numerals to count them, use a different form of mathematical expressions to analyze them. Whatever approach they have, they must reach the same conclusions and forecasts in terms of where the two stars will be concerning one another at different times. The reality of two stars existing with certain properties and dynamic behavior concerning one another will always be the same in all worlds.

So numbers, their behavior, and their meanings are independent of the mathematical equations that govern their behavior. Prime numbers, π, φ, e, h, c, α, G... are going to be what they are, whether we exist or not. The concept of "Unreasonable effectiveness of mathematics," stated in 1960 by Eugene Wigner Nobel Laureate in Physics is:

"Mathematics is real, and it is discovered."

Fundamental constants in mathematics and physics are not derived; they seem to be the properties of and etched in the fabrics of the mathematical and physical world. They show up in most of our equations, in macro astronomical observations, micro quantum mechanical measurements, and they allow us the predictability, consistency, regularity, and continuity of understanding of the dynamics of our mathematical and physical world. What is so particular about these numbers? Are they hardcoded in the fabric structure of our universe? Are they the constant numbers needed for our universe to be the way it is? Were they required and set for the initial conditions for the big bang to start? Are they set and fine-tuned to be as they are? It seems like so[47].

---

[47] I recently came across an interesting YouTube video by "Neil Turok" from the Perimeter Institute for Theoretical Physics, entitled: "The Astonishing Simplicity of Everything". It tried to explain a unifying equation based on a generalized wave

Related Quotes

1. "There is for me powerful evidence that something is going on behind it all. It seems as though somebody has fine-tuned nature's

---

equation incorporating fundamental constants e, π, i, h, and G in nature. It is summarized as follows:

"The wave equation for general universal motion that incorporates all the known fundamental forces in nature acting throughout space-time on matter, includes very important transcendental mathematical constants e, π, i and physical constants h, and G. It is as follows;

$$\Psi = \int e^{\frac{2\pi i}{h}A}$$

And;

$$A = \int \left(\frac{R}{16\pi G} - \frac{F^2}{4} + \bar{\Psi}_i D\Psi - \lambda H \bar{\Psi}\Psi + |DH|^2 - V(H)\right)$$

Where Ψ is the general Shrodinger wave function for matter calculated through the Feynman's integral, and A is a representation for all the possible known forces acting on matter, in space-time (also see footnote 44). To simplify A, we can write it as;

$A = \int (\,$ + Einstein Theory of Gravity effects shown through $\frac{R}{16\pi G}$

− Maxwell-Yang-Mills, Electromagnetic, Weak & Strong effects through $\frac{F^2}{4}$

+ Dirac's equation effects through $\bar{\Psi}_i D\Psi$

− Kobayashi-Maskawa-Yukawa particle effects through $\lambda H \bar{\Psi}\Psi$

+ Higgs effects through $|DH|^2$)

− Lagrange effects through $V(H)$)

Does this mean that our physical universe is somehow connected through these constant numbers? Are these constant numbers the common factors equilibrating the forces of nature? It certainly looks very strange that we can find and calculate certain constant numbers in mathematics and physics, based on our theoretical understanding and tested by experiments, with such simple and beautiful inter-relationships. At the same time, these constants allow us to predict and understand the dynamics of our physical universe at the micro (quarks, strings, etc.) up to macro (astronomical, galactic, etc.) levels?

numbers to make the Universe. The impression of design is overwhelming."

Paul Davies (British astrophysicist)

2. "The laws [of physics] ... seem to be the product of exceedingly ingenious design. The universe must have a purpose."

Paul Davies (British astrophysicist)

3. "I would say the universe has a purpose. It's not there just somehow by chance."

Roger Penrose (mathematician and author)

4. "Astronomy leads us to a unique event, a universe which was created out of nothing, one with the very delicate balance needed to provide exactly the conditions required to permit life, and one which has an underlying (one might say 'supernatural') plan." I would say the universe has a purpose. It's not there just somehow by chance."

Arno Penzias (Nobel prize in physics)

Questions

1. **Show the relationship between the metallic ratios φ4 and φ1.**

   Solution:

   We know;
   $$\varphi_1 = \frac{1 \pm \sqrt{5}}{2}$$

   $$2\varphi_1 = 1 \pm \sqrt{5}$$

   $$\varphi_4 = \frac{4 \pm \sqrt{20}}{2} = 2 \pm \sqrt{5}$$

   Then only for the positive signs;

   $$\varphi_4 = 2\varphi_1 \pm 1$$

2. **Simplify the ratio of two consecutive metallic ratios φn+1 and φn.**

   Solution:

   We know;

$$\varphi_n = \frac{n \pm \sqrt{n^2 + 4}}{2}$$

Then;

$$\frac{\varphi_{n+1}}{\varphi_n} = \frac{\frac{(n+1) \pm \sqrt{(n+1)^2 + 4}}{2}}{\frac{n \pm \sqrt{n^2 + 4}}{2}}$$

And we get;

$$\frac{\varphi_{n+1}}{\varphi_n} = \frac{(n+1) \pm \sqrt{(n+1)^2 + 4}}{n \pm \sqrt{n^2 + 4}}$$

3. **Knowing that the speed of the electromagnetic field or wave in vacuum c is inversely related to the magnetic constant $\mu 0$ and the electric constant $\varepsilon 0$, as follows:**

$$c = \frac{1}{\sqrt{\mu_0 \varepsilon_0}}$$

Interpret this speed in terms of the meaning of two fundamental physical constants.

Solution:

The magnetic constant $\mu 0$ is interpreted as the permeability or the ***allowance of movement property***, and the electric constant $\varepsilon 0$ is interpreted as the permittivity or the ***resistance of movement property*** of fields in free space or vacuum. The constant speed of the electromagnetic field or wave in vacuum results from the interaction of the two said opposing properties concerning the movements of fields in free space.

4. Knowing that the rest mass of an electron can be calculated from the Plank constant h, the square of the fine structure constant a, the speed of the electromagnetic field or wave (light) in vacuum c, and the Rydberg constant as follows:

$$m_e = \frac{2R_\infty h}{c\alpha^2}$$

Incorporate the metallic constant φ2 (Silver) in this equation.

Solution:

We know;

$$\varphi_2 = \sqrt{2} + 1$$

$$2 = (\varphi_2 - 1)^2$$

And we get;

$$m_e = \frac{R_\infty h}{c} \left(\frac{\varphi_2 - 1}{\alpha}\right)^2$$

5. Using the results from question 9 in the previous section as follows:

$$Plank\ Sphere\ Volume = \frac{\pi}{6}\left(\frac{HG}{c^3}\right)^{\frac{3}{2}} = 2.2105\ldots \times 10^{-105}\ m^3$$

Determine how many Plank spheres would fit into an electron shaped like a sphere with a diameter of $1.0 \ldots \times 10^{-18}\ m$.

Solution:

$$Electron\ Sphere\ Volume = \frac{\pi}{6}(D)^3 = 5.2360\ldots \times 10^{-55}\ m^3$$

$$Plank\ Spheres\ in\ an\ Electron\ Sphere = 2.3690\ldots \times 10^{50}$$

6. Using the results from question 9 in the previous section as follows:

$$Plank\ Sphere\ Volume = \frac{\pi}{6} \left(\frac{HG}{c^3}\right)^{\frac{3}{2}} = 2.2105\ldots \times 10^{-105}\ m^3$$

Determine how many Plank spheres would fit into a proton shaped like a sphere with a diameter of $1.6828\ldots \times 10^{-15}\ m$.

Solution:

$$Proton\ Sphere\ Volume = \frac{\pi}{6}(D)^3 = 2.4951\ldots \times 10^{-45}\ m^3$$
$$Plank\ Spheres\ in\ a\ Proton\ Sphere = 1.1287\ldots \times 10^{60}$$

7. What is the equivalent energy for the rest mass of an electron defined as:

$$m_e = \frac{2R_\infty h}{c\alpha^2}$$

Solution:

We know;

$$E_e = m_e c^2$$

Then;

$$E_e = \frac{2R_\infty h}{c\alpha^2} c^2 = \frac{2R_\infty hc}{\alpha^2}$$

8. Show the relationship of the fine structure constant α in terms of e (charge), h, μ0, ε0.

Solution:

We know;

$$\alpha = \frac{e^2}{2\varepsilon_0 ch}$$

$$c = \frac{1}{\sqrt{\mu_0 \varepsilon_0}}$$

If we substitute for c, we get;

$$\alpha = \frac{e^2}{2h}\sqrt{\frac{\mu_0}{\varepsilon_0}}$$

9. Assume that the universal equation stated in the footnote, presented by "Neil Turok" from the Perimeter Institute for Theoretical Physics, is correct. What type of wave dynamics and behavior would the following wave equation integrated from -∞ to +∞ in space (vector r), represent:

$$\Psi = \int_{-\infty}^{+\infty} e^{\frac{2\pi i}{h}A} d\bar{r}$$

Solution:

The equation for $e^{\frac{2\pi i}{h}A}$ can be simplified as $e^{(\alpha A)i}$ where i is a complex number equal to $\sqrt{-1}$ and α equals to the constant $\frac{2\pi}{h}$. As stated in footnote 22, we then can expand as follows:

$$e^{(\alpha A)i} = \cos(\alpha A) + i\sin(\alpha A)$$

The above equation shows an oscillating, sinusoidal, or increasing and decreasing circular dynamic behavior (due to the factor 2π and i) at the quantum mechanical levels (due to the factor h), for the function A. Function A represents a combination of all possible known forces acting in an oscillating, sinusoidal, or increasing and decreasing circular dynamic manner in space-time. When integrated over all possible states from -∞ to +∞ space, it will show the wave equation for all possible states when all possible forces are acting. It is very interesting to see the fundamental transcendental numbers e, π, and i, physical constants h, G, with all the possible known forces being incorporated in this universal oscillating, circular, sinusoidal equation. Is this how the universal dynamics are? Maybe, Einstein's dream came true!?

10. **As in the previous problem, assume that the universal equation is correct. What kind of assumption has been made concerning the behavior of matter in the universe? Does matter have a particle or wavelike behavior?**

Solution:

We have;

$$\Psi = \int_{-\infty}^{+\infty} e^{\frac{2\pi i}{h} A} \, d\bar{r}$$

Which shows a wavelike solution, based on the wave function Ψ and it is calculated using the forces incorporated in function A. Therefore, everything in the universe will be affected by all the known forces with different degrees through function A, in a wavelike, oscillating, and sinusoidal manner.

APPENDIX A

Continued Fractional Analysis

When you express a real number as the sum of its integer and the reciprocal of another number (the fractional part) iteratively, and you write this other number as the sum of a new integer part and another reciprocal, and so on you are using the Continued Fractional method of Analysis. When you have a finite continued fraction, the fractional term terminates after specific steps. When you get infinite continued fractions, you are generally dealing with irrational numbers, and we can say that all infinite and periodic continued fractions represent irrational numbers. The integers created through this iterative process are called the coefficients of the continued fraction.

For the irrational number π we get:

$$\pi = 3 + \cfrac{1}{7 + \cfrac{1}{15 + \cfrac{1}{1 + \cfrac{1}{292 + \cfrac{1}{\ldots}}}}}$$

We do not observe any particular patterns in the integer coefficients in the denominator, which continues until infinity. In Fractional Analysis terms the results of integer coefficients are shown as

[3;7,15,1,292,1,1,1,2,1,...]. If there are any patterns in the coefficients, the periodicity is specified. For the number π, there is no periodicity. In the case of $\sqrt{2}$, an irrational or none ending fractional number that is equal to 1.4142... we get:

$$\sqrt{2} = 1 + \cfrac{1}{2 + \cfrac{1}{2 + \cfrac{1}{2 + \cfrac{1}{2 + \cfrac{1}{...}}}}}$$

The number 2 pattern shows up in the denominator and continues until infinity with a periodicity of 2, repeating forever. The Fractional Analysis can be shown to result in a set of integer coefficients such as [1;2,2,2,2,2,2,2,2,2,...].

The continued fractional analysis is a powerful tool in dynamical systems analysis, mainly when dealing with Chaos dynamics and Mandelbrot or Fractal structures.

Fine-Tuning Principle

The principle stating that there are universal structures and ratios of important dimensionless physical and mathematical constants (such as the mass ratio of proton to electron, gravitational constant, plank constant, expansion rate or mass density of the universe, speed of light, the values of π, e, and golden ratio to name a few) that are delicately designed and finely tuned, without which the world as we know would have never been stable and existed. If a slight change or deviation is made to any of these finely tuned constants or structures, the world will fall apart. It is deduced that for the world to be and exist as it is, it had to be finely tuned and designed and caused by a grand designer or God.

Plank Scales

Max Plank, the great German physicist, utilized fascinating analysis and mathematical derivations, concluded that there are minimum scales for Length, Time, Mass, Charge, and Temperature in nature that we can observe and calculate. These scales were derived from five critical constants of nature; the Speed of light "C," used in Relativity, Reduced Planks constant "H or Plank constant divided by two times π," used in Quantum Mechanics, Gravitational constant "G," used in Gravitational theories, Coulomb constant "Ke," used in Electricity and Magnetism, and Boltzmann constant "Kb," Used in statistical behavior of gases and laws of Thermodynamics. These constant are universal and constant through time. If they increase or decrease, the calculated scales will also change (remember the Fine-Tuning Principle). The equations and values calculated for the five important Plank scales are as follows:

Minimum Plank Length = $\sqrt{\dfrac{H*G}{C^3}}$ (1.616×10^{-35}) Meters

Minimum Plank Time = $\sqrt{\dfrac{H*G}{C^5}}$ (5.391×10^{-44}) Seconds

Minimum Plank Mas = $\sqrt{\dfrac{H*C}{G}}$ (2.176×10^{-8}) Kilograms

Minimum Plank Charge = $\sqrt{\dfrac{H*C}{Ke}}$ (1.875×10^{-18}) Coulomb

Minimum Plank Temperature = $\sqrt{\dfrac{H*C^5}{G*Kb^2}}$ ($1.417 \times 10^{+32}$) Kelvin

ZF Set Theory

It is a revised version of the set theory put forward by Ernst Zermelo, Abraham Fraenkel, in the early 20th century, using certain axioms and logical deductions to resolve Russell's paradox proposed on the classical set theories of the time. It is also known as the Axiomatic Set Theory. This theory is based on eight crucial axioms (a more straightforward version has five axioms). Without trying to go through the mathematical rigor and notations, a brief description and importance for each axiom are as follows:

1. **Axiom of Extensionality**
   It provides the logical conditions to allow the equality of two sets if their elements are the same, regardless of the order of the set elements.
2. **Axiom of Foundation**
   It provides the logical conditions for not allowing any set to be a member of itself. It helps to resolve Russell's proposed paradoxes.
3. **Axiom of Specification**
   It provides the logical conditions for the creation of new sets from the members of each set. It allows for more diverse sets to be created.
4. **Axiom of Pairing**
   It provides the logical conditions for the creation of new sets through union and pairing of several other sets without reference to their elements. It also allows for more diverse sets to be created.
5. **Axiom of Union**
   It provides the logical conditions for the creation of new sets through the union of elements of several sets. It allows for more diverse sets to be created.
6. **Axiom of Replacement**
   It provides the logical conditions for the creation of sets resulting from the output of functions affecting other sets. It allows for more diverse sets to be created.
7. **Axiom of Power**
   It provides the logical conditions for the creation of power sets. It allows for more diverse sets, which include powers to be created.
8. **Axiom of Infinity**
   It provides the logical conditions for the creation of infinite sets and the natural number system. Using this axiom, it is proven that natural numbers are a set of all number sets. It allows for more diverse subsets to be created.

Some of the ZF axioms can be derived from more basic ones and, therefore, cannot be considered as independent axioms. As can be seen, most axioms allow the creation of new sets, and some others limit the creation of paradox causing sets.

Irrational Numbers

Includes all numbers that cannot be written as the ratios of two integer numbers (not including zero as the denominator) and will have decimals that will never end. It includes numbers such as $\pi = 3.1415....$, Euler's constant or $e = 2.7182...$, Apery's constant or $\zeta(3) = 1.2020...$, Golden ration or $\varphi = 1.6180...$, $\sqrt{2} = 1.4142...$, $\sqrt{3} = 1.7320...$, etc., or $Ir = \{\pi, e, \zeta(3), \varphi, \sqrt{2}, \sqrt{3}, ...\}$.

ABOUT THE AUTHOR

Shahin A. Shayan is a global investment & risk management consultant. By the end of 2016, he was the chairman of the board of Hoda International Financial Engineering Company, a private global investment banking/corporate finance advisory operation. Currently, he advises on management structures to create socially responsible economic enterprises, complex financing setups, investments, restructuring, enterprise-wide risk management frameworks, corporate valuation, and privatization issues related to the US and the Middle Eastern companies.

Initially, he worked as a young research scientist at NASA's Jet Propulsion Laboratories in Pasadena, CA on Molecular, and Ion-Ion Scattering problems. He later gained extensive financial experience in US firms such as Goldman Sachs & Co. in New York City, Security Pacific Merchant Bank, and First Interstate Bancorp in Los Angeles. He later gained valuable operational experience as an executive running investment operations related to volatile and uncertain developing environments in the Middle East, the challenge which he found very appealing. He innovatively structured and started two valuable Orphan Funds by raising funds from the official Capital Markets in the Middle East, helping 20,000+ orphans in the regions.

Born in Teaneck, New Jersey, he spent his teenage years growing up in Tehran. He speaks, reads, and writes Farsi fluently. Between the years 1976-84, he earned his BA in Quantum Chemistry, BS in Chemical Engineering, MS in Chemical Engineering, and an MBA from Columbia University in the city of New York. While working, he received his Doctorate Degree in Business Administration from California Coast University in 2014. Throughout years he has completed five specialized Executive Management Programs at Harvard Business School in Boston, MA.

He has been an active university lecturer and Executive Management Program speaker for almost 25 years. He has written six books, many scholarly papers on a wide range of topics related to Financial Engineering, Corporate Finance, Islamic Finance, Investment & Risk Management, Corporate Governance, Corporate Social Responsibility, and structuring Joint Social & Economic Enterprises, to name a few.

An Ivy League All-Star Soccer Team member for four years, he was drafted professionally and played Semi-pro Soccer in New York City leagues for ten years. He was the only Ivy League soccer player to be chosen as the Most Valuable offensive US Soccer Player (MVP) in the year 1979 at the prestigious annual US Senior Bowl Soccer tournament in Tampa, Florida. He was also chosen on the 2016 Columbia University All-time Hall of Fame Soccer team. After 40 years, he still maintains the all-time joint assists and goal-scoring records at Columbia University soccer.

www.ingramcontent.com/pod-product-compliance
Lightning Source LLC
Chambersburg PA
CBHW070240220526
45465CB00004B/1464